John Stevens:
A Pearl Harbor Survivors Story

John Stevens:
A Pearl Harbor Survivors Story

By John R. Stevens

Editors: Don and Denise Downey

WWII Veterans History Fund
2021

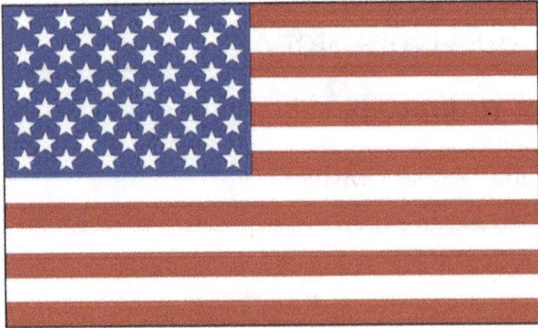

First Printing: 2021

ISBN 978-1-63010-029-2

WWII Veterans History Fund
San Ramon, CA

www.dondowney.com

Dedication

To my lovely wife Jodi.

Without your support and patience, I would have never achieved my goal of living to be *100 years old.*

Thank you.

Contents

Acknowledgements

I would like to thank my fellow Veterans, especially the local Marines, San Francisco Korean War Memorial Foundation members and USS San Francisco Memorial Association members.

San Francisco Chronicle Newspaper Article 04JUN2021
by Sam Whiting

The Korean conflict is known as "the Forgotten War," but one veteran who never forgot it was Lt. Col. John R. Stevens, U.S. Marine Corps.

Lingering pain from frostbitten toes were a regular reminder of his part in the Battle of Chosin Reservoir, fought in minus-40 degree weather in 1950.

Stevens earned a Bronze Star in Korea, a war that technically never ended. After his retirement, he got involved in another war seemingly without end — the long slog to get a Korean War memorial built in San Francisco.

He spent seven years fundraising, working alone in a windowless office on Van Ness Avenue, before the $4 million monument overlooking San Francisco Bay was finally dedicated in 2016.

Every year after, Stevens marked the anniversary of the start of the war, on June 25, by giving a speech at a ceremony there and laying a wreath on the memorial wall. This year's ceremony will be more somber than the others: It will be in memory of Stevens himself. He died May 25 at his home in San Francisco. He was 100.

"The Korean War Memorial would never have happened without John," said Don Reid, who also served in Korea and co-founded the memorial foundation with Stevens. "He was a true Marine, the total package. He stood for pride, commitment, dedication, virtue, honesty, loyalty and patriotism."

Stevens was also on the ground during the bombing of Pearl Harbor and the eventual defeat of Japanese combatants at Okinawa. For his valor under fire, he received his first Bronze Star.

"John was at Pearl Harbor, the Battle of Okinawa, the Pusan Perimeter, Inchon Landing, the liberation of Seoul and the Chosin Reservoir," said Gerard Parker, executive director of the Korean War Memorial Foundation. "From the beginning of World War II through the first crucial year of the Korean war, John had a knack for turning up in these key battles."

Through it all, Stevens was soft-spoken, modest and dry, using no more words than the setting required. Once asked by an interviewer what it was like to be surrounded by 100,000 enemy troops at Chosin, he responded, "Lots of targets."

After 23 years in the Marines, he went on to work 35 years in business, mostly in telecommunications and information technology companies. He started as a systems engineer at IBM and went on to start four companies. One of these, Centex, went public in 1987 and became a case study at Harvard business school, as an example of "how to run a successful IPO," said his son, John R. (Steve) Stevens II of Lafayette.

"My father was an inspirational leader who touched many individuals over his long life," said Stevens, who is CEO at Jopari Solutions, a Concord health care IT company that was another of his father's ideas. "He was kind, ethical, humble, and he always finished what he started. He said he was going to live to be 100 and he finished that, too. He made it by a month."

A Pearl Harbor Survivors Story

John Richard Stevens was born April 22, 1921, in Butte, Mont., where he grew up. The gloom of the Great Depression hit by the time he was a teenager, and after working stints as a baker, lumber jack and firefighter, he caught a train to Salt Lake City, hoping to join the U.S. Navy for the steady pay. He failed the eye exam and was headed to the train home when a recruiter for the Marine Corps spotted him.

"A man in a blue uniform with a red stripe on the leg caught me by the arm," he recalled in an interview for the Korean War Memorial newsletter. "I didn't know what the Marine Corps was, but I didn't want to go back to Butte. The rest is history."

Assigned to the 1st Defense Battalion, Stevens was a 20-year-old sergeant on duty at Pearl Harbor when Japanese bombers made their surprise raid on Dec. 7, 1941. After three years in World War II, he'd made captain, assigned to the 1st Marine Division Signal Company, where he worked with Navajo code talkers.

On April 1, 1945, Stevens arrived on the island of Okinawa as part of the largest amphibious landing of the war in the Pacific. Two months of fighting cost more than 12,000 American lives but the island stronghold was wrested from the Japanese.

Five years later, North Korea invaded South Korea and Stevens was the commanding officer of Able Company, 1st Battalion, 5th Marines. Assigned to the Pusan perimeter, his unit came under attack at the Battle of Obong-Ni, nicknamed "No Name Ridge." Stevens led his men in a counterattack and again earned the Bronze Star. "His outstanding display of personal courage, devotion to duty and leadership was an inspiration to his command," reads the Bronze Star citation.

He was pulled from the front lines to prepare for the secret amphibious landing at Inchon on Sept. 15, 1950. Arriving by landing craft, Stevens managed to lead his rifle company off the beach and over the seawall under steady fire.

Stevens went on to lead troops as they fought in the streets of Seoul. He survived that and the Chosin Reservoir Campaign, and was finally sent home in late November 1950. In two wars in the Pacific, he'd been in six of the biggest battles, many in command of a rifle company, which made him a preferred target. He earned 14 service awards and medals and got out with only three frozen toes on his right foot.

Those toes bothered him the rest of his life, but not as much as the fact that there was not a memorial to the Korean conflict in the city where he'd made his home and built his business career.

One day in 2009, Stevens was having lunch at the Marines Memorial Club with Reid and Man J. Kim, a Korean American restaurateur, when Stevens asked in a voice barely above a whisper, "Why don't we have a memorial here?" No one had a good answer and by the time they walked out to Sutter Street, the Korean War Memorial Foundation had been started.

They already had an office, the one where Stevens did consulting work. He cleared off his desk and got started. Reid was the treasurer, and Stevens was the secretary. They knew they needed a bigger name to serve as president, and for that they recruited former Rep. Paul N. "Pete" McCloskey, who was awarded the Navy Cross for his heroism leading a platoon in six bayonet charges in Korea. Stevens tracked down McCloskey on his tractor at his farm in Rumsey (Yolo County).

"Pete joined the memorial board as president because of John," said Reid, a retired banker at Wells Fargo. The Presidio Trust offered them a site across from the San Francisco

National Cemetery. The promontory has a view of the Golden Gate, through which Marines sailed when returning from the war before disembarking at Fort Mason.

The Korean War Memorial, which received major funding from the government of South Korea, was dedicated on Aug. 1, 2016, and Stevens was back at his desk the next morning. Now that the memorial was in place, he had to get people to visit and he had to continue raising funds for its upkeep. He kept that office until his death and was working there until a few months ago.

The memorial includes images of the war laser-etched onto a granite face. The dominant image depicts a platoon in Stevens's command going over the Inchon Wall. The platoon leader fell on a grenade to protect his Marines and Stevens made sure that he received the Medal of Honor posthumously.

Last year there was no anniversary ceremony because of the COVID-19 pandemic, so Stevens went alone to lay a wreath at the wall. This year the anniversary will return at 11 a.m. on June 25, with the public invited. Stevens's widow, Jody, and Reid will place a wreath at the wall, in honor of all who served and sacrificed in the Forgotten War, especially Stevens.

"John is gone," said executive director Parker, "but the memorial he built will last forever."

Survivors include his wife of 47 years, Joanne (Jody) Stevens of San Francisco; daughters, Carole Anne Clark of Great Falls, Mont., and Sherry Wilson of Colfax; sons, Mitch Stevens of Benicia and Steve Stevens of Lafayette; seven grandchildren; and nine great grandchildren.

Sam Whiting is a San Francisco Chronicle staff writer.
Email: swhit-ing@sfchronicle.com Twitter:@samwhitingsf

From San Francisco Chronicle online article: "John Stevens, decorated combat Marine who led the long battle to get a Korean War Memorial built, is dead at 100"
[https://www.sfchronicle.com/local/article/John-Stevens-decorated-combat-Marine-who-led-the-16223703.php#photo-21076847]

Start Time: 00:00:00
End Time: 01:44:22

QF (Female Interviewer)
QM (Male Interviewer)
JS (John Stevens)

(discussion about audio, set up)

JS: …first defense which I also had a searchlight battery with 60-inch search-lights. The theory was that the searchlight would pick up aircraft at night. And part of the battery was giant hearing horns and an operator would track on the plane some. And when they thought that they had the plane locked in, the searchlight would come in and then elec-tronically they would tell the three-engine aircraft what the speed and direction of the plane was so that it could fire. This was great in theory. We didn't have radar and it only worked though at night.

I was assigned to the searchlight battery initially as a wireman with the job of hooking up the sound-powered telephones to the various pieces of it. Before Pearl Harbor came I was transferred from that job, promoted and transferred to the machine gun group. But while we were still in the States in 1940, I was detached as an enlisted umpire for a ma-neuver on San Clemente Island. I was assigned to an officer who was the officer umpire. The six marines were stationed at San Diego and they were the main force on this maneuver. My job was to communicate using signal flags. We did not have handheld radios. As a mat-ter of fact, there was one radio in the entire battalion at headquarters and it was on that one man pumped a generator and cranked it while the other man used Morse code to communi-cate.

We didn't have landing craft at that time and we were not on the USS Nevada. And we landed in regular longboats, boats that were designed not to run up on the beach and dis-charge troops. I was wearing green uniform and the boat could only go up so close to the shore. We jumped off over the side up to water to our chest and then go ashore; wet, cold, and miserable. I just remembered that maneuver. Another maneuver that we had was out on Kearny Mesa. And this was before Kearny Mesa became built up. And we were laying down off the backend of a 6 x truck. And we were laying it from a spindle that had axles sticking out on each side and the wire rolled off of this. At that time we didn't have the utili-ties that they have today. We had baggy blue coveralls was our field uniform. And that spindle caught in the crotch of my coveralls and threw me down on the deck and they just rolled the wire up, which weighed about 300 pounds, landed on top of me. It broke a couple of ribs, broke my arm. So, I spent some time at Balboa Navy Hospital, but that all went by.

In February of 1941, the 1st Defense Battalion was loaded aboard the USS En-terprise and sailed to Pearl Harbor. When we arrived at Pearl Harbor -- in February it's warm, we were sent over to the navy yard and into the barracks that Dawn (05:45 sp?) so carefully marked out.

We immediately sent some people to Johnston Island, to Palmyra Island, and to Wake Island. In June of 1941, I was sent out to Midway Island to allow us another person, another communicator to intake back to Pearl Harbor for a month of R&R and I came back to Pearl Harbor. While I was out there they dispatched a major part of the battalion to Wake Island and I missed being on that detail by not being there. So, when December the 7th came there was only the headquarters and remnants and each one of the different units still at Pearl Harbor. While we had the antiaircraft battery there, the ammunition was in Una Luna Le (06:58 sp?), which is about two hours away. I was in bed at the time the attack started. And of course, I woke up and got out quick to see what's going on.

QF: So, what did you hear first?

JS: Well, I could hear bombs.

QF: So, you were asleep?

JS: I was asleep and the bombs woke me.

QF: And then you heard explosions?

JS: Yes. So, I woke up and got partially dressed and went outside. And I saw what was happening. They were dropping their bombs and torpedoes on the ships and then they were coming back and striking our area or the navy yard area. So, I got dressed and got my World War I tin hat, my World War I '03 rifle, my World War I gas mask and went on outside. I didn't have any ammunition, but we had guns and by this time I had been transferred from the searchlights to the machine gun group. I had been promoted to sergeant by now. Promotions were fast during that time. My primary job at that time was to lay wire to the different prearranged machine gun positions around the navy yard, one of which was down by the naval hospital.

They were already bringing in the dead bodies from the ships and from the harbor. By the time I got down to the naval hospital, they were stacking them up like -- a terrible sight. But after we finished laying our wire to our different locations, we then stood by to do anything. We had rumors of the Japanese landing on the north part of the island, that the water supplies had been poisoned, that the local Japanese were going to be attacking us; everything, all kinds of dumb stuff like that. But the attack was over in a very short time and then we had all these rumors. At twilight a flight of planes came in and everybody on the island opened up on them. They thought it was Japanese. It was a flight of planes from one of the aircraft carriers that was coming into Ford Island and out of six only two landed, which is pretty sad. But everybody had ammunition at that time. Everybody was ready to shoot and they all shot. That was kind of the end of the Pearl Harbor episode. In mid-December they loaded up what was left of the 1st Defense Battalion on a four-stacker, this tin can or destroyer, and we ____(10:28).

QF: Do you remember the name of the ship?

JS: I don't remember it. Davenport seems a likely name. It was a very, very small ship and we slept on the deck. There were no quarters for us. We got to Palmyra. Palmyra is a beautiful palm-covered atoll and a string of some 52 small islands around a lagoon. And there's one entrance into the lagoon that medium-sized ships can come in.

We had been building an airstrip on the largest of the islands, the main island, and that's where we had the barracks. And civilians had been building the airstrip and the CVs came in to complete it. And it was used as a ferry layover for planes going to the South Pacific. And at that time we did not have radar and the ships or the airplanes, many of them were lost. Most of the time we were covered by clouds and every day regardless we had a rainstorm. And all of our water ____(11:56) water we collected from the roofs of the buildings. We had no fresh water except the rainwater. They had plenty. While I was there I was the -- now I had been promoted to the Battalion Wire Chief and I laid a cable connected all of the islands back to the headquarters. At one point I had to place a cable in the water. The water between these islands was usually knee deep or so, so it wasn't a big problem, but it was a problem using hot lead close to water. I still remember that. It was the craziest. You'd never done it.

In August of that year I had now been promoted to Master Sergeant and then the Battalion Com Chief. I was offered a commission in the Marine Corps Reserve and I told the Battalion Commander I wanted to be a regular Marine. I didn't want that, and he said okay. And a couple of days later he called me back in and said, "Look, I accepted a reserve commission during World War I" and he had been called back in for World War II. And he says, "Now I'm a colonel." He said, "Don't be foolish. Take this commission." And I did. And the first thing the Marine Corps did was to transfer me from communications to be in charge of the newly arrived .40mm gun battery that nobody knew anything about and I knew even knew less about, but I moved out of my barracks into a dugout and I lost $3 in pay because it had longevity in my enlisted pay. So, the Marine Corps then sent me to Pearl Harbor to go fire control school so I could learn how to manage the .40mm gun battery.

As soon as I continued to fire control school the Marine Corps says, "We need a communications officer in Midway." So, I'm off to Midway again where I spent about six months at Sand Island. I worked as the Marines Communication Officer during the day and then at night I worked as a Navy Coding Officer. In those days we did not have very sophisticated coding. As a matter of fact, we have very unsophisticated, hard-to-use coding equipment.

I was then transferred from Midway back to Pearl Harbor and I flew back in the belly of a B-25 bomber who had ____(15:20) and I said why go now? Back up. I got back to Pearl Harbor and was then put aboard what used to be a banana boat, a very small boat, and it must've taken 10 or 12 days to get back to San Diego and I had _____(15:48). And I was assigned to the San Diego recruit training depot as the Base Communications Officer. And while I was there for six or seven months I was then sent to electronics engineering school at Great Lakes for a year of electronics training.

I should go back to Palmyra. In the summer of Palmyra we got our first radar, a CSR-270. And the CSR-270 was a very non-sensitive radar and it would identify clouds as planes. So, we had general quarters every day and we also had rain. However, we had radar. The Midway Battle took place in June of 1942 and I was in Palmyra at the time; when I went out to Midway that was all gone. However, we had lots of air alerts there as well, but we didn't have any action. The electronics engineering school at Great Lakes was one year and it was probably the best training I've ever had. It was a very small class, about 13 people including the lieutenant from the Turkish Navy. There were two other Marines and the rest were naval officers. When I completed that, I was sent to Camp Pendleton to take over the electronics school at Camp Pendleton. Before I got there the Marine Corps had decided I should become a line officer. I graduated at the second top of my class, so it wasn't because I failed to do my business.

At that time, the Marine Corps had one rifle regiment in a division and three understrength rifle battalions, two companies per battalion. And each company had one certain strength. They were very small. I was assigned to the combat service group and I became a company commander after, and then a battalion commander. And then when the 1st Marine Division reorganized in 1949, the colonel who had been in command of the 1st Combat Service Group was then given command of the 5th Marines. And he took myself and another officer with him. At that time I became the Commanding Officer of Able Company 1st Battalion 5th Marines. And we trained, and trained, and trained. Lieutenant Lopez, who is that man there, was the Company Executive Officer when I joined Able Company. So, he had an opportunity to train me because I had never been in an infantry company before.

In May of 1950, Lieutenant Lopez received orders to a school he wanted to go to on the East Coast, so he left the company. And at the time the Korean War broke out, we had -- see that top picture? That's a picture of the company in March. We had 105 men and we needed 205 _____(20:35) company. So, Lopez left the company. When the Korean War broke out, _____(20:49) shortly. The Marine Corps formed the 1st Marine Brigade. The war broke out the 25th of June and the 13th of July. The 1st Marine Brigade was made up of the 5th Regiment with six companies, three battalions, and the 1st Marine Air Group. They were 913 Air Group. It was a composite brigade and we had an artillery battalion and a motor transport company, self-contained unit. We had something like six days to intake enough people to come up to strength to issue weapons, to write wills, to relocate our families. Families -- some lived aboard base and they had to go someplace. I lived aboard base. My wife had to go someplace and embark.

So, on the 13th of July we left San Diego and arrived in Pusan. Initially we were supposed to go to Japan and prepare for an amphibious landing, but en route the situation got so bad in Korea. The North Koreans had pushed the soft greens in what part of the

Army was there down to a pretty close ring around Pusan on the tip of the Korean Peninsula. So, we were committed directly into Pusan. We were issued ammunition and we were committed the next day to combat, put to war. When we landed there, there was a feeling of fear. People were really afraid because from the 25th of June to the 2nd of August they had taken almost all of the South Korean Peninsula. They were literally unstoppable. The Army couldn't stop them. But finally, the Eighth Army had enough troops in there and with the brigade to hold them at this perimeter. While were there we fought three battles that were critical to the outcome. That first one was the North Korea had a breakthrough towards Masan and we were thrown in there and we pushed them back 30 miles. While we did that, they had another breakthrough on the Naktong River. So, they pulled us out. The Army held. And we were by train and truck moved over to the Naktong _____ (24:05) where we fought for three days, a very bloody three days, and pushed them back across the Naktong River.

On the morning that we threw them off of Obomini Ridge (24:19 sp?), I had 65 men standing. The rest of them were walking wounded, killed, or wounded. Now, the majority of the walking wounded came back within a day. Then we went back to an area for rest and recreation and to get ready for something else because MacArthur wanted to do a landing and cut off the North Koreans. We didn't know that. But why were we there? We were called back in the early part of September to the Naktong where they had broken through again and pushed the Army back again, and we retook the same ground that we had before. On the night of 6th September they pulled us off, took us back to Pusan where we were then given replacements and prepared for an amphibious landing. Now, among those replacements was Lieutenant Lopez who had fought his way back to get back to his company.

Now, my experience up to then was that if a platoon leader could last through one good firefight, he had a chance then of surviving. And at the time I had one deck sergeant as a platoon leader. I had one second lieutenant as a platoon leader and I assigned Lieutenant Lopez as a platoon leader, but for the inshore landing his platoon, a reserve platoon, so that the other two platoons by noon knew what they were doing. Now, normally before an amphibious landing, any landing, you have a reversal. For example, in World War II before we landed at Okinawa, the 1st Marine Division had a practice landing on Guadalcanal. And there we found out we had problems with boats, boats moved out. They weren't fixed. Tractors broke down. They weren't fixed. You find those things out. We didn't have time to do that at Incheon. Furthermore, we didn't have a beach that we were going to land on at Incheon. We're going to land a wall where there's a 30-foot tide. And we had to land exactly at the top of that tide.

These ladders were built aboard ship so that we could get out of the landing craft and get over the seawall. Now, the lack of practiced landing, my third platoon was with Sergeant McMillan. His boat broke down. He was in the left flank. My second platoon leader on the right flank, landed successfully. And his objective, we were landing right in front of a tall cliff, a proletary (27:47 sp?). And it overlooked the beach in Mandai (27:54 sp?) Hill had to be taken before the main body could come to shore. So, my second platoon, whose objective was a brewery (28:03 sp?) beyond that, moved towards their objective. The third platoon without their platoon leader was in a crevice just beyond what you see there. A hole had been blown into the wall and they were being held up by a machine gun. Lieutenant Lopez in his boat landed before the platoon leader of this other boat was able to land and his people started piling in there and we have a real problem.

Lieutenant Lopez wanted to take out this machine gun. So, he stood up shortly after there and unarmed a grenade by trying to throw it into the machine gun nets. And the machine gun caught him right across here. He had an armed hand grenade in his hand. He had people all around him. So, he's going down and he pulls the grenade underneath him. It absorbs the shock of the grenade, but saves all the people around him. In the meantime, the third platoon leader finally gets landed. The second platoon leader, I called him to come back because we needed help on the other side. He observes that the back of this cliff is a slope like this. So, he went up the slope and captured 16 North Koreans at the top, eliminated that problem. I sent the company executive officer over to take charge of the problem on the left where Lopez had been killed. And he took out the machine gun net and we were able to move in. In 30 minutes we had secured the beach, our beach, Red Beach. They were to fire off a flare. In that 30 minutes though I had 8 killed and 28 wounded, so it wasn't a cheap landing.

Oftentimes, Incheon is ____(30:21) as a not very expensive landing. Well, one-fourth of the total division casualties were in my company, so it wasn't cheap for me. But we then tied down -- it's dark of course by now and we settled down. And at high tide the ELSTs come in. And those guys thought there was a firefight going on and they opened up with their guns into us.

QF: Why would they think that?

JS: People get very nervous the first time in combat. And we had people the first night in Pusan who thought they saw something and fired. So, people the first time in combat are nervous and think they see things. Well, anyhow, we got them to shut it off. From Incheon we moved into Seoul, towards Seoul. We took the airport. We crossed the Han River. And on the way into Seoul the second platoon leader who had survived -- he was the only original platoon leader left of the original crew. We were taking many prisoners and collecting weapons. We were destroying weapons. And he picked up a Russian automatic rifle and swung it against a telephone pole to break the stock off so it couldn't be used. Well, the weapon fires on the forward movement of the boat, picks up a round, and fires. And the shock of it hitting that released the boat, put a bullet in the chamber, fired the bullet, and went right through his thigh. It took him out.

You know, bad things happen to good people. But we went on to the Han River and we crossed the Han River and into Seoul, and then recaptured Seoul. After we recaptured Seoul we were pulled out, put back aboard the same ship that we had come out from the States, the USS Henrico, and went around the southern tip of -- went to Incheon this time, got aboard ship. We sailed around the tip of the peninsula to Wonsan, which is on the east coast of North Korea. The North Koreans had mined the harbor ____(33:24). It was one of the major harbors of North Korea. So, we spent four days back and forth, back and forth while the minesweepers cleared out the harbor so we could go in. We were planning on an amphibious landing if necessary. However, by this time the Army had broken out of the Pusan Perimeter and had captured Wonsan, so they greeted us as we came ashore.

We then from there got aboard a train and went up to Anhong (34:04 sp?) and we got aboard trucks and started up towards the Chosin Reservoir, eventually getting to the Chosin Reservoir. Now, this is in October, late October. By now it's getting into November and it's starting to get cold and we still have our summer gear. In August in South Korea most of the time it's over 100 degrees, and at one time there were more people in sickbay from heat exhaustion than from battle. Now, as it turned around, same college, same stuff. So, now we're getting people with frostbite. We eventually did get something. The mistake that we made at this point, we should have stopped when we crossed the 38th Parallel because we had accomplished our basic mission, which saved South Korea. Instead, MacArthur wanted to go all the way to the Yamo (35:13 sp?) and that caused two more years of war, thousands of casualties than necessary.

In late November MacArthur was convinced that there were no Chinese, they weren't coming in, even though in early November I captured two Chinese soldiers on a patrol. He didn't see them. They just didn't exist. So we'll be home by Christmas, folks. By this time I was the only company grade officer, captain, or lieutenant of the original brigade that was still in the organization. The rest of them had been killed or wounded. So, somebody said, "Get him out of here." So, I was aired back out of the Chosin Reservoir area, flown back to Tokyo and then on back to the States.

When I came back to the States I wanted to visit some of the wives of the senior people that I had known. It was a small organization, so we knew our families. After two calls on wives, I stopped. There was great resentment that I came back and their husbands didn't. I didn't realize that. It was just sad. I initially was assigned after leave to the Concord Naval Weapons Depot as the Barracks Communications Officer. Remember, I was promoted very rapidly while I was enlisted. I went from private to master technical sergeant in less than three years. And I went to captain in less than two years, and I was a captain for seven years. By this time I had been promoted to major and I stayed at Concord for about two months and then was transferred to become the Commanding Officer of the Alameda Naval Supply Depot, and I spent the better part of two years there.

I was then assigned to Paris Island with two other Korean officers to screen recruits for OCS. And they had a backlog of 5,000 recruits that had tested to be eligible for Officer Candidate School. So, we did nothing for months except for test and interview, test and interview, test and interview. We finally got the backlog cleaned up. I asked the commanding general if I could have a meaningful job. The job I got was the Assisting G3 and the Senior Officer ____(38:38) court martial to try bad DIs. It was not a nice job. But shortly thereafter I was transferred to the Basics School of Quantico where I was put in charge of the weapons training group for training newly admitted second lieutenants. I did that for two years and then was transferred to Amphibious Warfare School for a year. I went back to Basics School this time as the S1 and Headquarters Battalion Commander. And from there I was transferred to Second Marine Air Wing. The Marine Corps had a program where four lieutenant colonels were sent to for air wings, border stations, and for ____(39:32) were sent to more Marine divisions, cross-training. So, I spent two years at Cherry Point.

After two years at Cherry Point the Marine Corps had a program that if you paid your way and you had the sufficient credits you could go to college. And I had more than enough credits for a degree and I elected to go to the University of Maryland where I finished a school year, graduated, and it was a degree in military science. I was then transferred to Kanoa (40:17 sp?), Hawaii as the Ground G3 Officer for a year, and then the Executive Officer of the 4th Marine Regiment, and then finally the Commanding Officer of the 1st Battalion 4th Marines. Now, of all the jobs I've had in the Marine Corps, being a company commander of any company was the very best because when you're in training everybody is looking over your shoulder from the Division G3 to the Regimental Commander; all these guys are watching you. In combat nobody. You don't see anybody. But it's a good job. You're really responsible. After I retired from Kanoa in 1962, I went to work for IBM. I worked there for seven years and then left to start a series of small computer communication companies. I finally retired from that in 2001. That's my story.

QF: Do you remember where you were when you heard the Second World War was over?

JS: I was on Okinawa getting ready to go to Japan. We were rehabbing Okinawa. And, you know, they talk about tornadoes here. Well, they have typhoons out there and those typhoons are so strong that we had one in early October that drove a destroyer up on the beach and it nearly completely wiped out our camp. But anyhow, I was there. People had their rifles and shooting. There were people happy. We were particularly happy because we didn't have to go to Japan, but instead we went to China.

QF: What did you do there?

JS: We went there with the job of repatriating the Japanese. The 1st Marine Division went to North China. The 6th Marine Division went to South China. And I spent six or seven months in North China. We were at Beidahu, which is just south of the Manchurian Border, so I was able to see the Chinese Wall inset Shanghai Kwon (43:00 sp?) on the border at China Manchuria.

QF: How did you like China?

JS: Well, at that time it was an extremely poor country. And we'd go out and lay telephone wire during the day. Our job was connecting stations from Pataho to Taishan to Tackotinko (43:22 sp?), about 300 miles. And we'd lay wire in the daytime and the Chinese would come out at night and steal it. So, we finally had to go to radio. It was interesting. I was young at the time. I probably didn't appreciate it as much as I did when I back there in '89 and went back to the same area. It had quite changed.

QF: What did you think about the atomic bomb?

JS: Well, I thought the atomic bomb was a very good thing because if it hadn't been for the atomic bomb, we would have slaughtered hundreds and hundreds of thousands of Japanese and Americans, and it saved them. Now, it's true it also killed a lot of people. That's true. But you know? The Japanese and the Germans had no compunction about killing anybody and everybody. That's hard.

QF: Do you have any particular combat stories?

JS: I told you mine.

QF: Just about those? Okay.

QM: And you do have two medals. Maybe you could mention those to us, please? Thank you.

JS: I was awarded a Bronze Star with a Combat V for my action at the first Naktong River Battle. And I was recommended for a Silver Star for action at Incheon, but the Army downgraded it to a Bronze Star. I sent for my records. You can send for your records and I did that a couple of years ago. And I found out that I had been also recommended for a Bronze Star during World War II, but it never was issued.

QM: And what was that for? What did your people write you up for that didn't quite --

JS: I don't know.

QM: Oh, very interesting.

JS: It was probably for China though.

QM: Thank you. And we would like to hear just some commander stories, just anything that comes to your mind good, bad, or whatever; when you were in charge. Thank you.

JS: When we fought the first Battle of the Naktong we had handheld radios. Our company radio net was handheld radios. We had an SUR-300 FM backpack radio for battalion net. Those handheld radios would hardly communicate across this room. They were useless. And after that battle the first thing I did when we got back close to an Army supply ____(46:48). I had my supplies sergeant go and steal enough SUR radios for all my platoons. From then on we had communications that I could control. I also had STO shielder halves (47:12 sp?) because we used up all of our sheilder halves in carrying out wounded and I almost got court martialed for that. General Craig, our brigade commander, saved me and the Army didn't appreciate the supply method ____ (47:32).

QM: Yes, sir. Well, that's Semper Fi is all I can say to that one. Now, what kind of guns were you trained on or what kind of guns did you like? Kind of a silly question, but --

JS: The company grade officers were armed with carbines, .30 Carbines. And the first time I tried to fire mine in anger or use it, it jammed. So, I got rid of that. One of my lieutenants had captured a US Colt .45 pistol from a North Korean officer, so he gave it to me and that's what I wore for the rest of the time. If I had a platoon a South Korean policeman attached to me during all the time I was in Korea, about 30 policemen all armed with carbines. And the lieutenant in charge, the platoon leader was my interpreter. And they were brave fighters. They were good men. We captured a bunch of North Koreans at one point in the Pusan Perimeter that I gave the prisoners to the lieutenant. I said, "Take these back to battalion headquarters where they can be interrogated." Okay. The battalion headquarters were maybe a couple of miles behind us. In about 30 minutes he's back. I said, "How come you could go so fast?" He shot them all. I said, "We don't do that." So, we got a lesson in human rights and you don't shoot everybody just because you don't like them.

QM: Well, that is something we'd like to maybe ask you about now. My dad had such positive experiences with the Filipino people and the Korean people and just felt like the Korean people just took back their country and how much they helped you guys. Did you get that same feeling?

JS: Oh, yeah. Many of the North Koreans became South Koreans because we evacuated over 100,000 just from Hungnam when we left there. And ____ (50:13) talks about that one ship, the Meredith, and what it had. But many other North Koreans fled to South Korea as well. And I know so many of those people and they're really South Koreans. The South Koreans can be just as cruel. They assassinated one of their presidents. Humans can be very cruel unfortunately. We try not to be, but that veneer of civilization is very thin.

QM: Yes, John. I would like to ask you another kind of serious question here because you're someone that has a lot of life experience and someone who has been brave and has been in combat I feel. How do you feel about war? And then extend that to the United States National Defense; just any thoughts that you could share with us? Thank you.

JS: Nobody likes war. As I think Sherman said, "War is hell." And that's true. And it's particularly hard on civilians. However, there are people who will not understand or change their way until they it's done by force. That means war. We've been fighting Islam for how many thousands of years? This is not a new war. This is a just a continuation of the Crusades and a religion that promotes killing nonbelievers. You can't negotiate with them. They're good negotiators. We'll lose. So, as much as I hate war, I don't know of an alternative to solving some of the problems.

QM: Yes. Well, and one of the most uplifting parts of this story has been after the war when peace comes. What do you guys do with yourselves? We remember those that we've lost because obviously they don't get to share in after the war, but do tell us a little bit more about your family life after the war and about your wife.

QF: Actually, I want to go back and I want to hear how you met your wife.

QM: Yes, sir.

JS: I've had three wives. Two of them are dead and one is alive. I've been married for 41 years to this wife. When I came back from overseas the first time in World War II, I went back to Butte and married a girl that I had dated in Butte before I went into the Marine Corps. She traveled with me during the 13 months I was in the States before going back overseas and became pregnant and had a baby while I was overseas. When I came back in 1946 from overseas we separated. And I later married my second wife and we were married for 23 years and we separated. And I met my current wife while I worked at IBM. As a matter of fact, when I joined IBM she was my trainer, one of my trainers. We knew each other for about six or seven years before we got married. But with my second wife I had one stepchild and two children, so I have a total of four children. One with one wife, two another wife and a stepchild.

QF: How many grandchildren?

JS: I have eight grandchildren and I have at least twelve great grandchildren.

QF: Did you say twelve?

JS: Yes.

QF: Wow.

JS: Well, my oldest son is 73 or 74. My youngest daughter is 64.

QM: Ah! So, that is so neat and just wonderful. We wanted to ask you before we maybe get into the computers, what about the Marines? Obviously you're a pretty tough Marine. You must be really proud to be a Marine. Do you go to reunions and things like that? Thank you.

JS: When I first got out of the Marine Corps I gave away most of my uniforms, which was a mistake. I totally separated and I now have a career, a civilian career. One of the things, I was competing at IBM with MBAs from Cal and Stanford. It's a tough place. And before when I was in the Marine Corps I used a much louder voice when I spoke. I learned to tone down my voice and generally conform to the civilian world.

QM: ...for interrupting, but especially a place like IBM. I mean I'm drawing a square with my hands because I'm thinking you guys are pretty square. Thank you.

JS: That's right, white shirts, black shorts, vests, hats. That was somewhat of a change. While I was at IBM I had a tour of duty down in Los Angeles. When I reported in to the Los Angeles office with a hat on and vest, they laughed at me. A different world down there; I spent two years down there and came back to San Francisco. But what was the rest of your question?

QF: Any reunions, Marine reunions?

JS: I didn't pay any attention to the Marine Corps really. Then I was on a business trip going back to Washington and there were two guys sitting up in front of me and they had Marine Corps hats on, the Marine Corp ____(57:47). I wondered, "What's that all about?" And I introduced myself and said, "What's going on here?" One of them was Colonel Mitchell Paige, Medal of Honor winner, and the other was Jim Forney (57:59 sp?). And they were going back to a reunion. So, Paige being the recruiter he was immediately recruited me. So, I started participating in 1st Marine Division activities and eventually was the vice president of the association and I've been on the board of directors pretty much since then. That was back about 1972 or so.

I got involved in the USS San Francisco because I saw a flyer and I didn't see the 1st Marine Division anyplace on it. And so, I contacted Chief Johnny (58:46 sp?) and I joined that board and helped them form a 501-C3. I wrote their bylaws and did all the things to make an organization. I spent quite a bit of time with them initially. And then we have a restaurant owner in San Francisco by the name of Man J. Kim, who owns maybe 13 or 14 restaurants, including Murray's Diner at the airports, Sears Fine Foods, and other restaurants. And starting about 10 years ago he started having a Korean Veterans appreciation dinner and I got involved in that. And then in 2009 he and I and another person were talking, another Korean vet. And we said, "We've got to have a Korean Veterans luncheon." And we talked to Marines Memorial and they said, "We'll give you a discounted price for everybody, but we feed the Korean Veterans free." So, we started having a Korean luncheon as close to September the 15th as possible. That led between the three of us to a memorial. So, we started working on the Korean War Memorial and that was in 2009. We started the luncheons earlier than that.

QM: We'll take a moment here and let's do talk about the memorial for a second. Why don't you give us like a two-minute sales pitch here and we'll just record that, John? Because we really -- this is just going to be incredible. This is the kind of first class thing that would be spectacular. I like it for two reasons. Number one, it's the Presidio. Obviously it's a place of honor, a place of burial. But on the other hand, you can look out and you can see where the guys are going out and back in their ships. It's kind of got that kind of significance too as kind of a harbor area to me, and plus they have the San Francisco Memorial nearby. But the Presidio is a special place, so we'd like to hear about that proposed memorial and your efforts there, John. Thanks.

JS: We spent several years looking for a site for this Korean War Memorial. And Don Reid, who is the treasurer of the foundation, was a sergeant in Korea in 1952. Don's neighbor is George Shultz's, who with is Carolyn Shultz, who is the protocol chief for the city. And she has connections with people at the Presidio and Don has connections with Dianne Feinstein, or his wife does. But with the help of these ladies we were finally hooked up with the Presidio. We asked Dean McCloskey, who is a retired Marine Corps colonel, a Navy Cross winner, a Korean Vet, a true hero, to be the president. And he also was instrumental in us finding a spot at the Presidio. The Executive Director of the Presidio was an intern in Congressman McCloskey's office at one time. So, it kind of fell together. But that didn't happen for a couple of years.

Once we found the site -- and they gave us the best possible site on the Presidio. It's across from the cemetery overlooking the Bay, overlooking the Golden Gate Bridge. It's at the corner of Sheridan and Lincoln, and Sheridan is probably the most traveled road or street in the Presidio. Lincoln is the most foot-traveled street because they're on the way to the Golden Gate Bridge. So, we have a site there. We're erecting a monument to commemorate the service of those who served in Korea. Now, there's over 2,000 Korean Veterans buried at the cemetery. And the Presidio was the base from which many were trained and sent to Korea and certainly many were returned, both alive and dead.

The memorial will consist of a one-third acre plot at the junction of Sheridan and Lincoln that will be beautifully landscaped and will have a long, curved black marble wall with laser-inscribed scenes of the Korean War, and of course a map showing the phases of the war. And there will be rocks from Korea. We already have rocks from the ____(1:04:37) China Wall sent by the Mayor of Incheon. There will be tiles along the Lincoln side of the memorial in honor of or memory of people who want to be remembered. There will be plaques for a more expensive cost, benches, and we have major donors who will be recognized on the entrance walls, on inside the entrance. The entrance will have a Bronze Medallion of the United Nations with all of the flags of the United Nations countries who served in the Korean War.

Now, the cost of doing this is about $3.4 million and that includes an endowment for maintenance of almost a half-a-million dollars. The Korean government has passed in their assembly a bill authorizing or pledging a million dollars, providing certain requirements are met. We have raised $2.1 million. So, we have about $400,000 more to raise. We have had tremendous support from Consul General Han of the Korean Consulate and all of his staff and the Korean community both in small organizations, schools, churches, and individuals. We just couldn't ask for better support.

QF: Do you have any kind of idea when?

JS: We're going to have groundbreaking the 5th of June this year.

QF: And how long do you think it will take to build, a year or two?

JS: According to the Presidio's plans, it's going to take us into 2016.

QF: Yeah, that's not bad. Wow, it's happening.

JS: It's happening. It's happening. This version is slightly modified. It doesn't show the curved wall.

QM: Well, let's maybe mention a few of your major contributors. You mentioned the Korean government. And I guess Quenton Kopp is involved, is he? I'm trying to think who.

JS: Pete McCloskey resigned a couple of months ago because of ill health. And he lives in Ramsey, which is a four-hour car ride. And Quenton Kopp has just been elected as the new president. Quenton Kopp, as you know, is a judge, former supervisor, former senator, and former president of the High-Speed Rail venture. Quenton Kopp is the president. Man J. Kim, who I mentioned earlier, is the vice president. And the secretary, John Reid, is the treasurer. Jerry Parker is the executive director and we have one other full-time employee who works here normally.

QM: And what's her name?

JS: Well, the person you met is -- she's gone. The person who replaced her is Eleanor Isaponta (1:08:15 sp?).

QM: Wonderful. I want to change the subject. We want to get into the computers and I have one final question. But I want to get into the computer thing. I want you to just tell me again just anything about the computer business. And then how in the world did you break away from IBM and start your own company? What was that like? And I want some real details on the company that you started. That just sounds like some incredible history of early computing with what's -- anyway, I'll let you tell what you were doing. Thank you.

JS: My last job at IBM was the National Account Manager at Fireman's Fund Insurance Company. I knew nothing about insurance when I took that job. However, the people, the executives at Fireman's went overboard to teach me the insurance business. When I joined them they were still using the 7000 series tape-driven computers and had a couple of 14-in-one-card computers. And this was the beginning of the 360 revolution. I was the IBM salesman that sold them the 360 series computers and some other specialized computers. But their data processing department, of course, programmed these computers. And the 360s were being programmed in COBOL (1:09:57 sp?). They had to have a conversion from the 7070s to the 360s; a huge conversion project, but that got done.

They had a high turnover. They would bring people in. They had a good training program, the data processing people. Then they'd go find a job with Southern Pacific or PG&E or one of one of the other large companies, and it was just going like this. So, after a couple of years out there I told the vice president in charge of that area that you ought to spinoff your data processing department into a company where the employees could have stock and be vested in the company, that way you're going to keep your employees. So, he thought it was a good idea and he took me to the chairman and we talked about it and the chairman thought it was a good idea. And he gave me $50,000 to -- or made me a consultant. I resigned from IBM. They made me a consultant, which was more than I was making at IBM. When I went to work for IBM I was paid $500 a month and as a consultant I was paid $5,000 a month. And my job was to develop the business plan to spin-off the data processing department. So, I didn't know anything about it; we were developing a business plan. I knew the business would be to develop insurance software both for them and other insurance companies.

I went to an accounting firm, one major accounting firm, and paid them to develop the business plan. That's where I spent the $50,000. I took it back and I made a presentation to the executives. Well, the chairman sat here and he sent over a life insurance company and a property and casualty insurance company. And the president of the property and casualty insurance company says, "There's no way that our employees are going to have stock in a company." Well, when I wrote the business plan and wrote the agreement between myself and the company, I wrote it so that if they didn't accept the business plan then it was my property and I could go with it. And so the chairman said, "Go with it." Now, of course, my first customer was the Fireman's Fund because we were converting them to an IBM property and casualty processing package on the 360s, which were the first tracked access systems that they had ever installed. So, that's how I got started in the computer software company business.

The company that's on _____(1:13:19) American Information Development, and that company is still alive today. The insurance company is really paperbound and to get an automobile policy, for example, the agent has to write out a request for a proposal to the companies. And then the company underwrites. He sends in the information for underwriting. They underwrite it and then they send in paper back with the proposal. And then the proposal is taken to the customer, if accepted, then back for a policy. And that policy gets written and delivered. It's pretty terrible. So, I did a study on the paper flow. I had done this study while I was at Fireman's Fund and decided that there was an opportunity to put communications into the agent's office to communicate to the companies. And then I had to find out how could I justify putting hardware into the agent's office. At that time there were no personal computers, but there were modems-based workstations. And there was a company called Cydcor (1:14:54 sp?) that had sold one of these systems.

So, we developed an application to send out to the agent where they could enter in policy information and transmit it over a dialup line to the company. We later changed them to lease lines. But that wasn't enough justification to have the hardware in the agent's office. And their biggest potential use of a computer was their accounting system. And for worker comp people, first report of injury, a claim can't be opened for a worker's comp claim until they receive the doctor's first report of injury and the agent can't send in his report or a claim and get paid for it until the company has the doctor's first report. So, that led us to the physician's office and to the claims office, so it kind of grew.

QM: And there's another whole world of paper at the physician's office, but sure. Go ahead, John.

JS: No, we worked it.

QM: Okay.

JS: We worked it.

QM: You weren't into that yet.

JS: We weren't ready to deal with that one yet.

QM: Exactly. But go ahead, sure.

JS: We were on the interface with the physician's office, but not to --

QM: To handle their work yet?

JS: No. So, this kind of grew in 1974 or '75. Continental Insurance was at that time one of the large insurance companies. They became a client of ours and they were interested in what we were doing. And on their board was the chairman of Q-tron Systems (1:17:11 sp?). And Q-tron Systems had built many a computer and they had at that time all the business that Bloomberg has today on PCs had on minicomputers with timesharing workstations in broker's offices. So, we made a deal to take my company over and provide customers and the product. Q-tron would provide many computers and we set up regional minicomputer processing centers. And Continental loaned us $100 million and we were going to go out and automate the agents. That was their interest. They wanted to have automation with the agents. And that went on for a couple of years. We went through that $100 million like you can't believe, but after about two years IBM came out with the PC. And I went to the board of directors and I said, "This PC can do everything in the agent's office with no lease line, no lease line cost, no terminal cost.

QM: And you can rent it as a terminal. But go ahead, yeah.

JS: That's correct.

QM: It does everything.

JS: It's going to kill the minicomputer or timesharing business. And I was right, it did. Q-tron (1:18:47 sp?) went out of business later. So, Continental bought us out and I went on to form another company. This time it was a communications company. A friend of mine, a neighbor, had -- AT&T had just broken up in 1984 and Pac Bell, Pacific Bell, could only sell service within its service areas. The long distance carriers could only sell long-distance. Well, my friend had found a tariff that allowed you to combine those two services into a group and sell that service, but there were two friends. One was the engineer/promoter. The other was a lawyer. And they didn't know how to start a company. So, I was footing free and I said I'll do this for you. So, we formed another company called Syntax Telecommunications and started implementing this program.

The process of doing it was to go out to a company like Intel, who has a lot of long-distance traffic, a lot of local traffic. And so you gave us your bills for the last six months and we would analyze those bills and tell you how much we would save you. Well, part of the deal was that the engineer had another engineer who was an expert at analyzing bills. So, he was doing this analyzation by hand, but it's very time-consuming to bring all this data up. So, we rented an erstwhile OCR machine, one of the very first that had been built, and used that to convert the telephone bills so that they could be analyzed. And then we'd go out and say, "We can save you this much money or this much. Sign on." And we had a hockey stick like this of revenue.

Well, we ran out of money real fast because we funded it was friends and people and we finally got a venture capitalist, who had a hire from AT&T who understood exactly what we were doing. So, they came in and gave us about $35 million. And then we looked for a --

QM: And this was when? Excuse me, John. This what year, do you remember approximately?

JS: This was about '85.

QM: Thank you. Wow.

JS: We hired the president of a telephone company down in New Mexico to take my job. I was the president and CEO and chairman. And he came in and he became the chairman and CEO and president. We trained him and he took the company and it did extremely well. We had offices all over the country and someplace it was voted the number one --

QM: The fastest growing public company and things like that, sure.

JS: We took it public and they had a secondary offering. So, we made a bunch of millionaires and it was a good experience.

QM: A little bit of your entrepreneurial spirit, huh?

JS: I formed another company after that. And that was given to me by one of the employees of that company.

A Pearl Harbor Survivors Story

QM: Well, anything else you want to talk about your business career? Or we can go on to -- let's talk about the free enterprise system. When I went to college in economics they taught me there's good and there's bad. If you have a wild -- what do they call that laissez faire, systems it'll get out of control? You overregulate. That's never good for business. And then you can go so far as a communist system, which to me just seems like another bunch of gangsters. But anyway, go ahead, John, anything about your thoughts about economics and starting businesses?

JS: Well, I'm a free enterprise man. The business that I started there, I was able to get a patent on the business system.

QM: Yeah, explain that. Thank you.

JS: It's a system you interface between medical service providers, payers, bill reviewers, all the people in the payment process. And we implemented that system, started it. Got venture capital financing again, about $30 million, and the PC bubble broke -- burst in 2001. The venture capitalists got scared and pulled their money out. We still had about $15 million in cash. Our biggest customer, Hartford Insurance Company, bought the company, bought us out. About five years after that my son, who had worked in the company, resurrected the company and now has about 12 million transactions a year ago ____(1:25:05). He has over a hundred employees and he's doing very well and has no venture capital and no money loaned to him.

QM: Well, that's something I wanted to lead to. There are people like your son and yourself that seem to be able to handle computers that just understand that it's all ones and zeros and has this ability to deal with these large millions of transactions and is comfortable with programming. Yet it seems like most of the population is not comfortable with those kinds of jobs. Have you seen that how most people really don't enjoy programming computers or don't really want to get involved in that dirty work, which again has made you a fortune?

JS: Well, I'm not a very good programmer. I was good at solving bugs in programs oddly enough. I did that, but it takes a certain aptitude to be a programmer and not everybody has that aptitude. For example, it takes a certain aptitude to be a radio operator. You have to have tone sense and I don't have any tone sense, but I score high on aptitude tests and intelligence tests.

QM: So, what did you bring to the table, John, your ability to?

JS: Organize. I provided an organization for the USS San Francisco for this organization and for these other organizations.

QM: Awesome. Now, I'm going to a totally off topic to a silly question. This is more like a question we ask the guys in the Navy for World War II. Did you cross the datelines or the equator and go through any of those silly ceremonies? And if you remember anything like that, tell us now.

JS: I went through one once. I only had to do it one time. I went through a cross many times; many, many times. And I've been around the world several times. But as I recall, I just got dunked and paddled.

QF: Do you remember what ship you were on?

JS: I don't remember. Wait a minute. I've got a certificate.

QM: Yeah, we're going to have to check your certificate or you're going to have to go through the ceremony again, John. So, take your time. So, yeah, maybe the next thing is why don't you take us on a tour of your office here and maybe I'll lift up a camera and --

QF: He's kind of hooked up.

QM: Yeah, so we can probably unhook him. Do you want to do that now, take a tour of the office and we'll talk about -- yeah. Why don't you go ahead and unclip him and we'll do that then.

QF: Excuse me.

QM: So, we're going to just let this one roll. I'll take the ____(1:28:28). Well, John, while we're moving here and switching cameras, I've got to say what a good job you did today. Don't you agree?

QF: Oh, yeah. He was awesome.

QM: Good job, John, very good. Very good interview, thanks.

JS: I'm about to lose my voice.

QM: You were great. And I think, Denise, it was good on the microphone, right?

QF: Uh-huh.

QM: Yeah, you're going to be fine with that microphone right there next to you. So, thank you for hanging in there. And I'll step on the other side of you if I could. Thank you. Excuse me for kind of getting behind your desk. So, why don't you just go ahead and show Denise whatever and I'll just relax here and video?

QF: This is the famous poster obviously.

QM: And we do want to show the plans for the memorial, although it's probably secret. But that will be spectacular and we'll be sure to put a link to the website.

QF: So, that picture up there, that was your --

JS: The top picture was my company at Camp Pendleton in March of 1950 before we went to Korea. The picture below that is my battalion in Kanoa (1:30:00 sp?) in 1962.

QF: That's quite a bit bigger.

JS: This great big picture up here, when I came back in 1943 the Marine Corps collected this group of people, they were all ex-enlisted men, and sent us through a special three-month class called Reserve Officer Training Class. They taught us -- at that time I was going on four years of service. And there were people who had much more service than me in there. And before we started the class, the colonel in charge of the school called us in and said, "I know that you're not going to like this, but do it. Just deal with it."

QF: So, which one are you?

JS: I'm at the top row, the extreme left, second person.

QF: Okay.

JS: My hat is crooked.

QF: Your hat is crooked, okay. You've got the crooked hat.

JS: And the picture below is a picture of the Taj Mahal of India.

QF: Where did you go there?

JS: We took a train trip in India a couple of years ago.

QF: That's pretty exciting.

QM: Well, excuse me, but I can see crossing the equator right here.

JS: And then down below I think --

QF: Naval Order of the United States?

JS: No, down below. It's the plank holder.

QF: Which ship are you the plank holder of?

JS: San Francisco.

QF: Oh, my gosh.

JS: I'm also a plank holder of the USS America.

QF: Wow! What's the fire brigade?

JS: That's what they called the 1st Marine Brigade in Korea because we were putting out fires. We were the only guys that were offensively-minded there unfortunately.

QF: Really?

JS: Yeah.

QF: You'd think fighting a war they all would be offensively-minded.

JS: We were just trying to hold them back.

QM: Now, how do you feel about the Chinese people after that experience?

JS: Well, the Chinese people are okay.

QM: Well, that was something though. I just hope we can educate everyone, but as you were saying, just millions of Chinese came across the border, right?

JS: There's hundreds of thousands.

QM: Hundreds of thousands. Thank you.

JS: If you ever tried to kill the 1st Marine Brigade -- we fortunately had a good division commander who disobeyed MacArthur's chief of staff, General Lawman (1:32:58 sp?).

QM: Well, I'll tell you the same thing I tell all the Cold War Era, and I guess Korea is kind of part of that Cold War, that since we really gave them a whipping there in Korea and Vietnam in my opinion, I think it did make the leaders in Moscow and Beijing think twice before trying something in Europe or against Japan.

JS: That's right. They stopped. They did stop. You know something? The United Nations hasn't taken an inch of effort towards these other countries that are being invaded by Russia. I don't understand that.

QF: ____(1:33:36). I don't either.

JS: They took Crimea. They're trying to take --

QF: The Ukraine.

JS: Out of Ukraine, out of Georgia.

QM: And we're not so naïve as to think that those aren't Russian troops training and shooting with those people and running all of that fancy equipment.

JS: Absolutely.

QF: So, is that you there?

JS: That young guy?

QF: Yeah.

JS: Yeah, that's me. That was 1962. That's one of ____(1:34:05).

QF: Oh, did you get a picture of that, Donald? That's when you retired?

JS: Yeah.

QF: When people retire they usually look older than that.

JS: (Laughs) I was 42.

QF: You still looked pretty good for 42.

JS: Well, I went in when I was 18.

QF: Yeah. So, what's the one just above crossing the equator, the Peruvian Natovian Rigues (1:34:34 sp?) or something -- probably one of those things from the Navy again?

JS: I think that's the 180th Parallel.

QM: The Golden Dragon?

JS: Yeah.

QF: Oh, you have the Golden Dragon. Wow.

JS: ____(1:34:52). I'll have to take it down and read it.

QF: I love these sketches here. So, this says "Operation Boomerang." It's kind of an unofficial certificate.

JS: What happened in 1961 before the start of the Vietnam War is that we loaded up the 1st Marine Brigade in Hawaii, told everybody that we were going to maneuvers in California, went over to the ammo depot, picked up a load of ammo, and then headed out towards Vietnam. So, we got one day off the coast and they pulled us back or we would have landed in ____(1:35:43) in 1961. So, the people who went on that trip got the operation.

QF: Do you know who drew that up for you, just somebody there that happened to be a pretty good artist?

JS: Well, the 1st Battalion 4th Marines was a pretty unique unit. They generated a lot of generals and they have a reunion periodically down in Las Vegas.

QM: Wonderful. Well, I guess maybe we'll wrap it up this session. But believe it or not, we want to come back again and talk with you sometime or we'll work out maybe somehow we'll cross paths. But just thank you so much, John. Thank you.

JS: Well, you're welcome. I appreciate the effort that you're putting into this. Am I going to get a copy of this?

QF: Oh, gosh. Yes, probably multiple copies.

QM: All right. Thank you one more time and I'll unclick the button here to wrap it up.

QF: You've got so much to see on your wall. Wait, what's this? What's that?

QM: So, here's how we're planning on getting you -- I won't make any promises that I can't keep. But we're going to do the video for you and then we're going to type out the transcript. So, it'll be like a book thing. Then we'll, I don't know, throw in whatever else we can. Oh, we're going to do a map. So that reminds me, Denise. Let's have him look at the map here and then show me anything on this map that --

JS: Well, this is probably a better map.

QM: -- that he remembers. Okay, yeah it actually might be. Why don't you do me a favor and maybe just record him showing you on the map like where he was, anything he remembers about being at Pearl anytime. It wouldn't even have to be the 7th.

QF: ____(1:37:38).

QM: Especially like which one --

QF: It's not focusing, Donald.

QM: Oh, you can focus it with the top one maybe. Is that focusing?

QF: It's still not happy.

QM: Yeah, keep clicking on it.

QF: It's sort of focused.

QM: Okay. Yeah, it's pretty close up for this lens. Okay, now we're recording.

JS: At Pearl Harbor on December 7th I was in this barracks here. I believe this is No. 10 Dry Dock. No, this is No. 10.

QM: Yeah, with the Pennsylvania?

JS: Yeah.

QF: And where was the hospital where you were trying to lay down the lines?

JS: The hospital I think is down around here.

QM: Right. Yeah, we're on that side exactly.

QF: Is that all you remember?

JS: We used to have a train running from Honolulu out to Pearl Harbor.

QF: So, where did that go? Do you remember where the tracks were?

JS: No, you can't see it here. This is an only picture because there's a runway here now. There's a runway 90 degrees from this _____(1:39:16).

QF: So, you've been back to Hawaii?

JS: Oh, yeah. We just sold a condo on Kauai that we had for about four years, but I've been going out to the Pearl Harbor Day _____(1:39:35) in Pearl.

QF: So, you kind of got like a yearly thing you do?

JS: Yeah, except this year I was sick in the hospital.

QF: Well, maybe you can make it next year then.

JS: Yep, I plan on it this year.

QF: Oh, yeah. Sorry. Did you like the new visitor center?

JS: Oh, yeah.

QF: We went out there the year they dedicated it.

QM: Sixty-ninth.

QF: Yeah, the sixty-ninth.

QM: Yeah. So, we go out to the Arizona and I've got my little video camera. And all of a sudden two survivors show up with a widow of a survivor. So, I just kind of started to get into this interview thing or this video thing. So, I said, "Okay, I interviewed them." I interviewed the second guy right there on the memorial, right? And then he said, "We've got to interview the widow." And I go, "Oh, sure of course because I'm even into the ladies almost as much as you guys, the wives and what all they've been through; their lives too, right?" So, I interview her and then she starts talking about, "Well, I met my husband while I was a nurse and he came in and got shot up -- was it at Pearl actually?

QF: Yeah.

QM: "Got a bullet in his heart, came into the hospital." And anyway, she fixed him up and somehow I guess he made a -- you tell the story. Thanks.

QF: He proposed to her as he was going into surgery and she said yes because she thought the surgery he was going to get was pretty iffy and it was not very likely he'd come out of it. So it's like, "Sure, yeah. I'll marry you, sure."

JS: That's funny.

QF: Anyway, he came out of it and I guess he expected her to keep her promise and she did.

QM: And then I guess eventually he got that bullet out of his heart and something about the hole in his heart or some story. She had a nice phrase of how she filled the hole in his heart or something like that where you just -- every time I hear it I start crying; tearing up a little. But it was just such a crazy time and everyone's, from what I can see, everyone's life just got totally turned upside down especially like if there's any kids, child survivors and all -- thanks, John.

JS: I'm going to move you for a minute.

QF: Okay. We'll start clearing your office here.

QM: Well, you can trust us.

JS: Do you need to use the restroom?

QF: I'm okay right now.

JS: Why don't you do -- I leave the door open.

QM: We could probably come with you here and take a little tinkle myself. I need to put my shoes on though. Where did they go?
(packing equipment)

QM: Okay.

QF: I don't even know the other power cord for the lights are.

QM: Did I take part of it in?

QF: Well, I unplugged that one.

QM: The other one is right here. Wait, here it is. It never made it to on it. It's all tangled up there.

(end of audio)

Start Time: 00:00
End Time: 35:42

QF (Female Interviewer)
QM (Male Interviewer)
JS (John Stevens)

JS: The last time we talked when I talked about coming back from, from Midway to the states I inadvertently left out what happened for the next several years. I, I moved into a different segment. But I would like to go back and -- when I came back to the states in 1943, I was initially assigned to Camp Elliott to Howie Med Smith's Communication Section as a Coding Officer. And for a month, myself and a, a large number of other Second Lieutenants did nothing but sit at electronic coding machines, coding and decoding messages. After that month, I was given temporary duty orders to work along with the sequence school in New Jersey. However, the class was not due to start for a month so myself and two other marine officers ended up in Philadelphia at the Quarter Masters Depot salvaging British Tank Radio -- radios from British tanks to be converted for use in the US. After a month of that, I attended the Fort Manos Sequence School. That was a six month course. Again, I was given temporary duty orders to Harvard Univ-- excuse me, to Quantico, Virginia to attend a special basic class for officers who had been commissioned in the field. And that portrait on the wall there, there is about 40 of us that were in that class, most of them the same rank as I or more. After three months of that, I was then again given temporary duty orders to Harvard University to the Naval Officer's Communication School. While I was there, I started naval communications. I was also the student present of the student regiment. And I was selected bases on my voice, my ability to call 5,000 to attention.

At the end of that three months, I received orders back to the first marine division on Pavuvu in the Russell Islands. So I loaded up my 1939 Ford Coupe and drove across from Cambridge to San Diego via Butte, Montana. Got aboard a boat at San Diego and went to Pavuvu. Pavuvu is part of the Russell Islands, the same as the Solomon Islands, not too far from Guadalcanal. The first marine division that I had just completed the Peleliu Campaign and I was among the first group of replacements coming in for that. After a while on Pavuvu, we started training and getting ready to land on Okinawa. Part of the preparation for the Okinawa landing was practice landings on Guadalcanal. We finally loaded up ship in late February of 44 and started steaming towards -- north towards Okinawa in Japan. On the way, we stopped at Ulysses to meet with the task total -- Naval Task Force that we would be part of the landing force. Ulysses is an act-all and it had -- a large act-all and it had ships as far as you could see in any direction. It was probably the largest congregation of naval ships ever -- and, and that includes up to date.

We were allowed to give the troops two hour liberty ashore at the beer garden for a couple of days rotating groups ashore. Then we re-embarked and started towards Okinawa. On April the 1st -- April Fools Day, we landed at Okinawa. The landing was pretty much unopposed and at the time I was with the division headquarters in Signal Company and my job was to support the Assistant Division Commander. The Signal Company was broken up into two echelons. One afforded echelon that went with the Commanding General and supported him and then the other echelon, which I was, supporting the Assistant Division Commander. We had to lay off shore there for quite a while and during that time, the Japanese kamikaze airplanes conducted raids. I was on a communications ship that had antennas sticking out all over it, so it was a target for the Japanese. And I was never so scared in my life because there is no place you can go aboard ship. You cannot dig a hole in the deck. It's very frightening. So everybody was glad when we finally got off the ship and went to shore, but there was as I said a unopposed -- freshly unopposed landing. The first marine division, when they landed, their objective was to go across the island and then to turn north. The other units would go south. Well the Japanese had very light resistance going north, so that in a very few days we had accomplished all of our objectives. And then we were thrown into the south because they were not moving.

While we were there, we eventually captured the island -- and I forget the date. It seems to be some time in June that we finally finished there. The first marine division moved to the north part of the island to prepare for the invasion of Japan. While we were there, we accepted the capture of a company of Japanese that had hidden out in the rough. And finally we understood the war was over, I missed learning that the atom bombs had been dropped and of course, the war was over. Instead of us preparing to go to Japan, we prepared to go to China because the Japanese held all of the coast of China, all of the infrastructure and most of the population. Our mission was to go to North China, land at Dakutanku (sp?) and then go to Peking, drop a regiment there and send -- leave one regiment in Tientsin and my regiment, the seventh marines to go Idaho Beach, which was several hundred miles north of where we had landed. Just adjacent to Shanghai one, which is on the border of China and Manchuria -- and this was the beginning of the Great Wall of China where it starts at the sea. Shanghai is a calling port, probably four or five thousand people live there. Idaho Beach was the prime resort area of North China and all of the high value people summered there. I was there in China until the spring of 1946, this is 1945 now, and then I was returned to the states where I became the Communication Officer for the recruit depot. And I think we covered it from there on in.

QF: And what did you do exactly in China?

JS: Our job -- the regiment's job was to protect the coal trains and the railroad that ran between Shanghai -- Shanghai one and Peking -- Tientsin and Peking. Tientsin was the coal mining district and everything ran on coal. The trains ran on coal, the power plants ran on coal. It, it was you know, their lifeblood. So we had guards on the trains, and we had guards on the bridge. The communists were in Manchuria at the time and while we were there, they came down across the border to fight Tulu-Ling's (sp?0 nationalist armies. And of course, they destroyed them. There were units surrendering to the communists by, by the thousands. It was terrible to watch. But the people in China at that time were so poor, the nationalists were very corrupt and did not take good care of the people. The communists promised them something, so of course the people would sympathize with the communists.

For example, we would lay telephone wire in the daytime, at the nighttime the communists -- the local people would come and cut it up and give it to the communists. So we had to rely completely on radio.

QF: Do you remember any particular stories, any excitement?

JS: One bit of excitement we had was in early October, we had a typhoon on Okinawa. The typhoon was so bad it blew a destroyer up on the beach. It was -- I think it's still the worst tropical storm or typhoon in history. And of course, it blew our camp completely gone.

QF: How did you survive?

JS: Just down low.

QF: How long --

JS: Yeah.

QF: Wow. Nobody was lost though?

JS: No, none of ours. The Navy probably had some causalities.

QF: How did you like China?

JS: Well China was a poor dirty place. The fields were just as bare as this table. It, it was -- it was a poor, poor country. I purchased some antique coins while we were there and some other antique items, but not very much. In 1989 we went back to China and went back to where I had been. It's unrecognizable. Those fields that were bare are now tall with vegetation. It just -- so much different.

QF: In a good way.

JS: Yeah. I had a abscessed tooth. We spent a month in China that time and traveled all over -- took a boat ride down the yellow river to Shanghai. But I had it -- I went -- I had to go to a hospital and at this time, tourists were not allowed to go by their selves. We had a guide all the time we were there -- a government guide. And the hospital diagnosed my problem and they charged me $3, which included the medicine.

QF: Could they take care of it for you?

JS: Yeah, they took care of it.

QF: Oh they did --

JS: Yeah.

QF: Oh. For $3?

JS: Yeah. Yeah.

QF: Good deal. And how did you like China then when you went back?

JS: Oh it was -- it was interesting. You know, and we were there -- we left there the night before the tenement square massacre. So while we were there, we saw all these students converging on Peking and I said I don't understand why you know, they're allowing it. Well, they didn't allow it. They finally cracked down on them. But the people were still on bicycles there. We went from there to Taiwan and they're on motor scooters in Taiwan. Then we went to Japan where they were driving cars. Give you an idea of the level of sophistication.

QF: Well so how did you like Japan?

JS: Oh it was fine.

QF: Did you do anything in particular there or just passing through?

JS: Well in 59 I spent a month, month there too. So --

QF: And how was it in 59?

JS: It was quite, quite modernized.

QF: Oh so by 59 --

JS: Yeah, yeah.

QF: Okay.

JS: Oh yeah.

QF: And how were the people? They --

JS: They, they -- well they treat you well. They treat you better than the French do. We went back to Okinawa also a couple of years ago and I couldn't find the place that we had -- we went. The principal battleground is now a, a tourist trap. Times have changed.

QF: So did -- what parts of Japan did you go see?

JS: Well the first time I went back in -- it was mainly Tokyo, Nagoya, Niko, and we went of course, up to the mountain, Mt. Fuji. The second time I went back, it was the only way we could get to Okinawa was to go through Japan. I guess the second time I went back was from Korea. When I was finally taken out of Korea I, I had to go through Japan to catch air back. I'm out of gas.

QF: No. Okay, we can stop.

JS: So --

QF: Well great. Do you have anything?

QM: I guess we probably would -- since we're here John, and I know you probably don't want to brag or you know, but you do have a couple of medals. Maybe you just want to talk about those or about any of your combat stuff or whatever you want to talk about really, even if you just want to talk about philosophy, we're happy. But --

JS: Anybody that serves in the service and particularly those who go overseas collect lots, lots of -- sorry.

[off-topic telephone conversation 17:51-19:54]

JS: I have been waiting for a call from him for some time. I'm sorry.

QM: Do you want to grab this again? Looks like you're good. Well wonderful John and are we doing okay here?

JS So you want to talk about --

QM: But I have -- yeah, since we're all the way down here, we're going to probably press you for a little extra story or two. But we wanted you to talk about what you want to talk about. You know, we could do an interview, but -- so we'll just relax and we kind of -- but yeah, go ahead John and just keep --

JS: Well I'll, I'll tell you about, about the -- about medals. During World War II of course, we had all of the usual medals. A couple of years ago I asked for my records from St. Louis. In going through them, I found that I had been recommended for a bronze star. I never knew that and it was never awarded, so I don't know. In Korea, I was awarded two bronze stars. One of them had -- was -- had been put in as a -- for a silver star, but the army downgraded it to a bronze star.

QF: Did they say why?

JS: No, it didn't say why. My Battalion Commander -- even after he retired was -- kept fighting to get that up back to the silver star, but he was never successful. Some of the medals that are awarded in combat are, are most well deserved and there's many people who -- most of the medals that are awarded are well deserved. And some people don't get medals that are deserved, but General McArthur would come over to Korea with a bag full of silver stars. And one day I was asked to send some people over to get a silver star from General McArthur, which I did. You know, I picked --

QF: Did they tell you who to send over or --

JS: No, no, no -- oh no, just pick, pick whoever. I could have gone over myself. But I thought it was kind of a lousy way to operate. This person here was a congressional medal of honor and it's Lieutenant Palermo Lopez. The fifth marines fought in the first marine brigade at -- in the Prousanne (sp?) perimeter. And we were a regiment that only had two companies per battalion instead of three and we were the -- we were the fire brigade to save, save the Prousanne perimeter. If we had not been successful in what we did at the Prousanne perimeter, we stopped the North Koreans through three breakthroughs of the Prousanne perimeter, we would have been pushed off of Korea and Korea would have been lost. The same troops that fought there were selected by McArthur to head the Inchon Landing. So exactly the same companies, A company and B company, C company and D company, E and F were the people who did the Inchon Landing. Now Lieutenant Lopez was not with us in the perimeter -- Prousanne perimeter because he had left the company to go to school, but in the first batch of replacements he got, he came with them. He was assigned to reserve platoon and his platoon came in last at the Inchon Landing. Where he had landed though, they were being held up by a machine gun nest and he elected to try and take that out with a hand grenade. He threw the hand grenade and as he had armed it to throw it, he was hit by a machine gun -- the machine gun and he fell on top of the grenade. He pulled it underneath him and saved all the people around him. The company that landed -- Able company, that landed at Inchon, had a fourth of the whole division's casualties. We had 8 killed and 28 wounded -- no entire first marine division who is only 32 killed -- there was a -- all over in 32 minutes.

QF: Pretty intense 32 minutes.

JS: 32 minutes. I was able to fire a flare indicating that we had secured the area. Well have you got enough?

QF: Have we got enough? Yeah.

QM: Yeah.

QF: Well thank you very much.

QM: I guess we'll let you off the hook as soon as -- just give us an overview the whole project here, what you want out of it and how you want to be remembered or anything we can do for you, maybe this more or less off the record, but yeah John just talk to us about the whole interview thing, anything about life and then we'll wrap it up. Sure John, thank you.

JS: Okay. For a number of years Manjai (sp?) Kim, a local restaurant owner has been holding a dinner for Korean veterans every June the 25th and they are doing it again this year. We became friends. A couple of other new Korean veterans helped me host a luncheon every year on September the 15th or thereabout at the Marine's Memorial where the Marine's Club gives a free lunch to this Korean veterans and a low price for accompanying non-veterans. And working with that group, we -- back in 2009 talked about having a memorial for the Korean War here. And then we started looking for a place to put it and we looked at Golden Gate Park and the National Park, we looked at the Presidio and we finally made this an organization in 2010 and applied for 501(c)(3) status, which we are now, and started seriously trying to get a site. With the help of Don Reed who is one of the members and a friend of Diane Feinstein's, we were able to have a hearing at the Presidio at the Presidio Trust. And during this time we also asked Pete McClusky (sp?) who is a Navy cross winner and a Korean War hero, former congressman to be the President of this organization and he accepted. The Presidio showed us four sites that we could have, one of them is directly across from the National Cemetery overlooking the Golden Gate Bridge and the Bay, and that was the site we accepted. So then we decided we wanted to find a sculpture to do a sculpture that would relate to the Korean War. And there is a sculpture that has taken this and made a statuette out of it that's awarded to the top graduating students of basic school at Quantico each year for each class. And we hired him to do -- be our sculpture. And then of course, we had to worry about raising money and our budget as we figured out to do this job would be about three and a half million dollars. And that provided for long term maintenance as well as all the other work. After a while, the sculpture and the Korean government and the Presidio didn't like his work, so we switched to a different form. The form now, the upper picture there is what we're doing and the Korean government has allocated a million dollars to us. Their assembly has passed it this past January -- approved it. We have raised in addition to that about 2.2 million and very shortly we will sign the documents with the Presidio that will allow us to proceed. Our plan is to have the groundbreaking in July and it will either be on the 11th, the 18th or the 25th. And the reason we don't do the exact date is that we have some high ranking officers from Korea coming and we don't have their schedule yet. But we expect that momentarily. As soon as you go, I am going to be calling the consulate.

QF: When they expect to be completed?

JS: In late this year or early next year.

QF: Oh --

JS: Yeah.

QF: They did that fast.

JS: We have done, done an awful lot of preliminary work.

QF: I was going to say it looks very beautiful too.

JS: It will -- be -- what we're doing is beautifying a spot that's now just weeds and trees. So --

QF: A sculpture --

JS: There's, there's going to be a -- at the corner, at the junction of the two streets there, there will be a fountain. The people who are walking to the Golden Gate Bridge who walk up Lincoln, that's the way they go to get to the bridge, Sheridan is the main way to get north and west out of the Presidio, the main road. So there is a lot of traffic there, which is good.

QF: Well great. Are you satisfied? Okay.

QM: Congratulations John. And we'll look forward to that and remembering, as they call it, the forgotten war and what all you guys did to basically fight communism and support democracy. That sounds kind of -- I'm not sure what the right word is, ____(ilistic 32:32) or something. But that's what we believe in -- that's what I believe in and you can kind of see how China and maybe some day North Korea, we don't know -- what a country -- will see that you know, a little freedom, a little capitalism and --

JS: Well sure there's already a little capitalism.

QM: Yeah. Yes sir.

JS: They're capitalists to begin with. North Korea is different.

QM: Let's wrap that up with -- I'd like to -- and this is serious. What do you think about North Korea? You have more insight than most Americans. What should America -- how should America treat the government of North Korea?

JS: I think they should just, just leave them alone.

QM: Yes.

JS: Yeah. They're not the same people, but neither are the ISIS people the same people in our context. When North Korea was the industrial part of Korea, when it was a whole country and South Korea was the bread basket and the two of them really are com-plimentary and should be one country, and it's too bad that they're not. And of course we took hundreds of thousands of North Koreans out of North Korea with us when we left. I have talked to people from North Korea. Pete McClusky last year -- well the year before last made a trip to North Korea and met one of the old generals that had fought in the war.

The Koreans want to -- they're going to -- yeah, absolutely. They had a Reunification Council here in San Francisco -- one of our daughters. And the Korean people are really industrious people, but they are also -- have a very high flashpoint. They, they are very emotional, very committed, but they are really dedicated to whatever they're going to do. If it's bad, they're dedicated to that. If it's good, they're dedicated to that. So --

QF: Well great, well thank you so much and --

JS: Well listen, thank you and I hope that you're receiving enough reimbursement from Chevron to, to take care of you guys.

QM: Yes we are.

QF: Yeah, we are doing really good.

QM: Yes we are finally getting a little money back out of those guys after I gave 24 hours a day for 27 years. You bet you. So we're grateful for a little Chevron input to get the transcripts, so yes sir.

(end of audio)

Oral history interview of LT. Col John Stevens
by Edean Saito (NPS)

INTERVIEWED ON September 15, 2017 by Edean Saito

TRANSCRIBED by Colin Cordy on January 7, 2018

USS ARIZONA MEMORIAL
PACIFIC HISTORIC PARKS
NATIONAL PARK SERVICE
ORAL HISTORY COLLECTION

E. Today is September 15, 2017. We are in the home of LT.Col John Stevens. And we will be recording his oral history as a part of our Rendezvous with History program. So Lieutenant Stevens, LT. Col Stevens, is it okay if we just call you John?

J. Yes please.

E. When did you join the Marines?

J. I joined the Marine Corps in September of 1939.

E. At the age of?

J. At the age of 18.

E. Why did you join the Marines?

J. I had served two years in the Civilian Conservation Corps. Times were tough. I was born and raised in Butte, Montana, a mining town. And when the depression hit, it hit that town very hard. After I got out of the CCCs I couldn't find a job. Butte was a Navy town. If people went into the services from Butte for some reason they'd join the Navy. So I signed up for the local Navy recruiter to go into the Navy. There was a six month waiting before you could be sent to Salt Lake City for a medical exam. Because they didn't have these facilities in Butte, then being inducted into the Navy. Finally I was called up, got on a train in the afternoon. Played poker all night, when it came time for my physical exam I passed everything except the eye exam. And I failed the eye exam, so at that time they were really very strict. Because they were taking a few people into the services they were very small. As I was very sadly dejectedly walking out of the examination room, a man in bright blue pants and red striped on the side called me and says. "Here's some chits. Go get a good night sleep and food chits and come back here tomorrow." I didn't know it but he was a Marine recruiter. I didn't even know there was a Marine Corps at this time. So I did as he said. I came back, same doctor examined me. I passed no problem and I was on my way to San Diego to boot camp. That was 1939. I went through boot camp fine. I didn't have much of a problem because of my previous experience in the CCCs. At boot camp they had tests to determine where you would go after you completed. The tall good looking guys were sent to sea school. They would go on ships as sea detachment guys. The guys who passed with good intelligence test would go to communications. And they had two tests there the first one was a tone test if you passed the tone test then you became a radio operator. And if you didn't pass then you became a telephone man. I'm tone deaf, heehee, so I didn't pass the radio, so I went to telephone school. After I completed that, I was assigned to an organization called the First Defense Battalion. This was an organization that had been organized in anticipation of war in the Pacific. Remember now we are getting to 1940.

J. And this organization had a searchlight battery because we didn't have radar. We had a sound locator battery, anti-air battery, coastal artillery guns, and a very small machine gun company. I trained with that company in the searchlight battery. My job was to provide telephone connections between the searchlights, the sound locators, and the three inch guns. And the theory was at night the searchlights would find the airplanes and the sound locators would find the airplanes coming in. That information would be tracked to the searchlights and they would flash on. The same information would go to the gun and they could shoot'um. And this was not a great theory but that's the way it was. In the day time it was visual.

I was with that organization until the battalion was loaded up on the USS Enterprise in February 1941 to go to Pearl Harbor. One of these battalions had already been placed on Midway Island. Our battalion was destined to be split up; part to go to Wake Island, part to Palmyra and part to Johnston, and a core group at Pearl Harbor to act as a rotation group for the people in the outer islands. In the middle of 1941, I was with the people who had a rotating tour on Midway. I went out to Midway, spent a couple of months there and came back. By this time promotions came pretty fast. By this time I'm a Corporal, in charge of two other Marines. I was transferred to the machine gun group. On December the 7th I was asleep in my barracks bed. I heard all this noise. The barracks I was in, was maybe 1000 yards from Number Ten dry-dock at Pearl harbor. The Marine barracks, the permanent Marine barracks, was on the far side of the parade ground. We were on the near side to the harbor. I got up, I looked around, and I saw these planes. The planes were coming in dropping their bombs, then coming in and strafing the rest of the area where we were. I recognize what it was. I went back in, got dressed with my World War I tin hat on and my World War I leggings, cartridge belt, and my 1903 rifle which is what we were armed with at that time.

E. Why were you issued World War 1 equipment?

J. The Marine Corps got the bottom of everything. We didn't get new rifles until August of 1942. We were still firing bolt action 1903 rifles. My job, with these other two marines, was to run wire to gun emplacements that had been predetermined. So we went down to the gun shed. Now we had three inch guns and 50 caliber guns. But there was no ammunition there for the three inch guns. It was at the ammo dump. It wasn't till 11 o'clock that day that we got anti-aircraft ammunition. All we had up to that time was small arms and 50 caliber. But I got my crew together, the machine guns were deployed, and we started laying wire. It was a very sad scene. One of our gun emplacements was down near the naval hospital, you know where that is?

E. Mhm.

J. And they were bringing the bodies out of the water and stacking them up like cordwood by the hospital. The attack itself was over in a very short time. It was just a couple of hours, not very long. In my group, we had several people wounded, none killed. And I wasn't touched. We spent the day refining our positions and listening to rumors, like "the Japanese have landed on the far side, the water supply is contaminated." You can imagine the stories that would be circulating. That evening a flight of planes, as you know the aircraft carriers were not in the harbor, fortunately they were all out, a flight from the Enterprise to land at Ford Island.

As they came in people thought they were Japanese. Every gun on the island opened up on them. Of the flight, one or two landed, the rest were shot down, you know before they could stop the shooting. In early January they loaded, you know, what was left of our battalion. They put us on a World War 1 destroyer, and we went on the Palmyra. Are you familiar with Palmyra?

E. I know where it's located.

J. Its 500 miles southwest of Pearl, just above the equator. Where it rains every day and there was no fresh water supply. But there's enough rain water to supply all the water you need.

E. So you had to capture your own drinking water?

J. Yes, we had barracks there and the barracks' roofs captured more than enough water. We had civilians there finishing up an airstrip. The purpose of the airstrip was to ferry planes from the US to Hawaii to Palmyra and on into the South Pacific. I stayed there, and by this time I'm being rapidly promoted Sergeant, Tech Sergeant, Staff Sergeant, finally Master Technical Sergeant, and I was in charge of battalion communications. One day the Battalion Commander called me in and asked me how would you like to be a Lieutenant? I asked "a regular Marine Corps lieutenant?" He said, "No, a Reserve Lieutenant." I say no thanks and I went about my business. About a month later he called me back in and said "let's talk about this." He said "In World War 1, I was promoted to a Reserve Second Lieutenant and here I am now a full Colonel." He said I should take this, so I did. So the first thing they did was move me out of my barracks into a dug-out, and put me in charge of a newly arrived 40mm gun battery. A new artillery piece, nobody knew anything about it, including me. So they said, "we"ll send you to fire control school at Pearl." So they flew me up to Pearl and I finished the school. And then Marine Corps said we need a communications officer on Midway. So forget about the school you just finished. You're now on route to Midway. So they flew me to Midway are you familiar with Midway at all?

E. mhm

J. You have Sand Island and Easter Island. I believe they had the airfield on Easter Island. Or one of the two. I was to become the communications officer for that island.

I was a communications officer for the Marine Corps during the day, and at night I was a naval coding officer. Working in their naval coding office. And I did that for about, I don't know, six months or so. By this time, I had a lot of time overseas. So I was sent home, back to the states to go to school. I was in the states for 13 months with orders to Camp LeJeune, North Carolina. Permanent orders. What happened was, I first spent some time in a coding office at Camp Elliot. Just doing nothing but sitting at a coding machine. Then I was sent to New Jersey to go to the officer's school at Fort Monmouth radio school. Now I had to learn radio code. And I did, among other things. After I finished that school, I was sent to another school, this time at Quantico to go to an officer's training school after all this time. By this time, I'm a First Lieutenant about to make Captain. I was then sent from there to Harvard University, to learn naval communications officer school. And I did that, but in the meantime my orders were still to Camp LeJeune. And I'm not getting to Camp LeJeune.

E. Did you ever make it to Camp LeJeune?

J. I have never been to Camp LeJeune. What happened was my last set of orders transferred me to the 1st Marine Division in Bavvo in the Solomon Islands.

E. Why?

J. Well the First Marine Division, well this was the ending of 44 and the beginning of 45. The 1st Division had fought at Guadalcanal, went to Australia for rehab. After 6 months in Australia, they went to Cape Cloister. Then they went to Palau and then they rehabbed on the Russell Islands. It's part of the Solomon Islands group. And that's where I joined them as they were coming off the Palau Campaign. While they were getting ready to go into Okinawa, that was our next step. We trained at Bavvo, that particular island we were on. And then did retraining and trainings of practice landings on Guadalcanal. Then went to Ulithi. Ever heard of Ulithi?

E. mmh no.

J. It's an atoll in the Pacific that's big enough to hold the entire Pacific fleet. We came into that, I was aboard a command ship. At this time, I was the Executive Officer of the First Signal Company. We, the command, was broken up into two groups: the forward group with the commanding general and a second group with the assistant commanding general. And I was with that group. We were in different command ships. But as we came in, as far as you could see, there were ships. We spent several days there staging then left to go to Okinawa where we landed 1 April 1944. The Okinawa Campaign was kind of unusual because the Japanese decided not to put it all on the beach, but to let us come ashore and fight them. But in the meantime, the suicide bombers from Japan were just pounding us. And when you're on a ship, there's no place to hide. And you can't dig a hole in that steel deck. We had that for a day or so before we landed, after the initial landing waves. Is this what you are looking for?

E. Yes this is perfect. We want to hear your story.

J. We landed in Okinawa, and the First Marine Division was very successful in what they were doing. The Marines were reaching their objectives in time or ahead of time. The Army was having a tough time. The Marines went north and the Army went south. The Marines moved fortunately fast, and we were told after securing the north to turn around and help the Army in the south. But as we were going there, we noticed that all the graves were embedded in hill sides. They didn't have graveyards with gravestones and we didn't know why. Later we found out why. But we had incessant rain. We lost vehicles in the mud. It was just terrible. We had to resupply troops at the front by airdrops because we couldn't truck stuff up to them. So lots of rain, lots and lots of rain. And as the battle ended, I think in late June, we rehabbed the 1st division on the Motobu peninsula in north Okinawa and prepared for the landing on Japan. In October we had a storm and now we know why they had the tombs in the hills. The storm like a category 5 hurricane. It blew a destroyer up on to the beach. It wiped out our camp. Fortunately for the people there, they were prepared for it. Of course you know they dropped the bombs. And that caused us not to go to Japan, but to load up in late October and go into China to take the Japanese and send them back to Japan. Disarm them and send them back. By this time I had been transferred to the 7th Marine Regiment as a regiment communications officer. The 7th Marines were the lead regiment going into north China. We had a headquarters group to make a scouting run into the land to find out where to bivouac troops, where to get food and water and all that stuff. It consisted of the commanding officer, the operations officer, and the communications officer and staff people. Once we got ashore, another storm came in, and we were cut off from the ship for a couple of days. So we lived off the land. But the Japanese when they surrendered, they surrendered. They had stacked all their arms and armories. They were perfect prisoners and very cooperative. When the Japanese took over the coast of China, which was really all they had, they ran the country and they ran the industry. When they handed over to the Chinese [downward hand motion], it went to pieces, because the Chinese were not prepared for it at that time. It took them time to catch up.

E. They didn't have the training.

J. That's right, absolutely yes. So we put the Japanese aboard ships and sent them back. Then are regiment was assigned to a town, to guard the railroad near the Manchurian border in north China, to the city to Tianjin, where we had landed, it was about close to 200 miles.

E. That's a long stretch.

J. So we had a company in Chinwing town, a company in Tangshan, and another company down at the end. And we would have troops riding the trains. And the Chinese Guerrillas at that time, the communists of course would try to attack the trains. The principle source of coal in China at that time was at Tangshan.

E. This is really interesting. Most of what we have learned is all about what happened at Pearl Harbor, but not very much about what happened in the rest of the Pacific war. So listening to you is wow. This is great! This is like new stuff for us. And this is our mission. It's not just Pearl Harbor, but the Pacific war. So thank you for sharing.

J. I didn't do much at Pearl Harbor as you can see.

E. That's fine, you did a lot in the Pacific war.

J. Our headquarters was at Padiho. Are you familiar with Padiho beach?

E. no

J. It was a resort area for rich Chinese. Twenty or thirty miles south of Chinwangtao. The 8th Root Army, the communists, were up in Manchuria, and they came down through this area. We watched the Nationalists, who we had armed and trained, give everything to the communists. [shakes head back and forth] They surrendered by divisions, and they didn't put up much of a fight. That's why the 8th Root Army, the communist army, could move as fast as they did. Now that's not true for all the Nationalists, but it was true for that particular area. I don't know why, so we watched that.

E. It must have been very disappointing.

J. Hum while I was there the regimental headquarters with yellow jaundice. Including the commanding officer, the [points to himself] the communications officer, everybody. And it was the sickest I have ever been, even worse than what I have now. I recovered and eventually I flew out and came back to the States. And I was assigned as the communications officer to the Marine Corps base recruit depot at San Diego. I did that for a while, and then I was sent to a naval communications school at Great Lakes. It was an electronics, naval electronics school, and it was a year long school. Now radar is now in the picture. I completed that school and was reassigned to a logistics group. At the 1st Marine Division at Camp Pendleton. I had a number of jobs there, and in late 1949, I was transferred, they reorganized the division. At that time the division had one regiment instead of three regiments. And two companies instead of three for each of the infantry battalions. And they moved the commanding officer of the logistics group.

And he became the commanding officer of the 5th Marines, and he took two officers with him: a colonel, a lieutenant colonel, and myself. At that time I became the commanding officer of Able Company, 1st Battalion, and 5th Marines, which made me a lot more familiar with warfare than I had before. So I joined the 5th Marines, and we did some extensive training. And in June of 1950 I was on leave in Hollywood. I got a telegram, "the North Koreans had attacked South Korea, return to base." So I returned to base, and we had six days to take in replacements, issue gear and figure out where our wives are going to live before we boarded a ship to go to Korea. So on the 14th of July we set sail for Korea and on the second of August, we landed in Pusan. Fear was in the air. By this time, the North Koreans had pushed the South Koreans all the way down to the pocket around Pusan at the southeastern corner of the peninsula. And the 5th Marines with their two companies were the only marines that were available. So we were put into combat immediately, plugging holes in the army's lines. We stopped the North Koreans on three major drives in August and early September. On one of those nights after we had recaptured the ground from the North Koreans, out of my 205 man company, I had 65 men standing. You know the rest had been killed or wounded. But a lot of them were able to come back; they were minor wounds, they came back for the next battle. We had the highest casualties of battles in the Korean War. Took place in the Pusan perimeter. On the night of 6 September, we just pushed the North Koreans back again across the Han River. We secured the line, and this was kind of their final try. We were pulled off the line and back into Pusan port, where we received replacements to bring us up to strength. And we were preparing to make a landing on some place on the west coast of Korea. On the 13th of September we boarded ships, incidentally the ship we boarded was the same ship that we had boarded when we came up from state-side, just a coincidence. One of the officers who joined us at that time, a Lieutenant Lopez, had been the company executive officer when I had joined the company. And he had really trained me, but he left the company in March or April to go to a school. When the war broke out, he asked to be transferred back to the company. So he joined us on the 6th of September at Pusan, just in time for the invasion; we were very short on officers. Many times there would only be two or three officers, while officer positions were being staffed by NCO's who did a good job. For the landing at Inchon, I had three platoons now. Before I only had two. I had a lieutenant, a seasoned lieutenant, in charge of one platoon who would take the right flank, a technical sergeant who would take this other platoon on the left flank.

I assigned Lieutenant Lopez to the third platoon in reserve. My experience was once they had a firefight behind them, they were less likely to be killed, giving them a better chance of survival. As luck had it, we got off our ship at 5:30 on the 15th of September. Excuse me, we got off at 3:30 in the afternoon, for a 5:30 assault.

Picture if you can, a beach, but not a beach, but a sea wall like this [makes an upward motion with his hands] with a 30 foot tide. And right behind the sea wall is Cemetery Hill, some other feet high, overlooking this. That's our objective. So as we go in, I see that we're receiving fire, so I call for an airstrike. I have an air controller with me, so I am able to do that. So we had a funny little airstrike as people were maybe 50 yards from the beach. The right platoon made their landing, well they had ladders to go over the sea wall. They could not go out the boats, they got off and they achieved their objectives, which was a factory behind the Cemetery Hill.

The 1st platoon we had on the left side. The platoon commander's boat broke down on the way in. We didn't have any rehearsal. If we had a rehearsal, that would have turned up. But we didn't have a chance for rehearsing. So he was out here [making circling motions with his hands] drifting around in the water. The assistant platoon leader, a tech sergeant, took the troops in. Now when they got to shore, there was a hole in the wall, and there was a machine gun nest right in front of them. So they could not go forward. They were suffering casualties. Now comes Lieutenant Lopez, with the third platoon right into the same place. Now we got a platoon and a half in this small area. Lieutenant Lopez sees the situation, arms a grenade, stands up to throw it into the pillbox/machine gun nest, and they catch him across the body. So he falls down with an armed grenade, which is going to go off with this bunch of people. And he pulls it in to his body and of course gets killed. He dies immediately. I land about this time, now this is only about 5 minutes into this action. Hearing what happened, I sent the executive officer, Lieutenant Uneges, over to that area to get them moving, because we had to take that hill because there are thousands behind us coming in. You know these things roll in. In the meantime, my second platoon leader on the right side captured his objective. He called me, and I said come back and help. And on his way back, he sees that the hill is a slope on the back. So he starts up the hill first. He captures a bunch of North Koreans at the base of the hill, and they didn't expect anyone coming from behind. He gets to the top of the hill. It's alive with people, and he captures all of those without firing a shot. He takes them down to the bottom of the hill, and now the hill is secured. And the people on the left side by now have broken out. And they had made contact with the other platoon. So we had captured our objectives at H plus 35. I was able to fire an amber star cluster grenade, signaling that we had captured our objective. By this time [wave his hands horizontally], the tide is all out, and we have a mud flat here. The LST supply ships start coming in or have come in. They're sitting on the mud, and they come with their guns firing, very chaotic. But we stayed there for the night, and the next day we moved on through the city of Inchon. Then through the Kimpo airfield, and by the way, by this time the rest of the division has landed, the 7th Marines and the 1st Marines, but they did not have a combat landing.

E. Because you had already paved the way.

J. The division suffered that landing that day. If I remember right, 21 killed and [long pause, 10 seconds], large number of wounded. But Able Company had 8 killed and 21 wounded. We suffered more there, the most of the whole division's casualties. And when people say it was an unopposed landing, it really, really hurts me. It was unopposed for the rest of them, but it was opposed for us. Eventually we get to Seoul, and we take Seoul. By this time the weather is starting to turn. In early October they pull us out of that area, puts us back aboard ships and take us to the bottom of Korea. Up to Onessen. We were supposed to make an amphibious assault landing. But the army got there before we made the landing. One reason was the North Koreans had mined the harbor. It took us five days while they demined. The minesweepers could sweep the harbor, so that we could come in. So we had an administrative landing. We then got on trains and went up to Hamhung or Hang Nam. And from there, we went up to the reservoir in stages. By this time the 8th Army had pushed up all the way up close to the Yalu River. The Chinese are telling him, don't do it. And in late October I was on a company-sized patrol, and I captured a couple of Chinese soldiers, and I sent them back to battalion. Then they went to regiment to division and finally back to MacArthur. And he says ahh there's no Chinese, there are no Chinese. While in fact the Chinese were there and in force. We scraped with them on and off throughout October. It started to get really cold now. In late November I am the only officer of the original brigade group, company grade officer that hasn't been killed or wounded. So the Division commander said get out. So they sent me home. I was very sorry because right after that the whole thing erupted, and the rest of the Chosan Reservoir battle took place. So I'm back in the States,

A Pearl Harbor Survivors Story

and I wanted to go and talk to the widows and wives of people I personally knew. After a couple of them, I stopped. They were very resentful that I was able to come back and their husbands didn't. I stopped that. But I was initially assigned, when I was in North Korea I got frost-bitten in my hands and feet. We didn't have good clothing initially, or good foot gear. I was assigned to Port Chicago and was being treated at the hospital at Mare Island. After a couple of months of that, I was finally okay, except permanent nerve damage. I was then assigned as the commanding officer of the Naval Supply Depot at Oakland, which is really in Alameda. I spent a year and a half there, and the Marine Corps decided that they needed to get more officers out of the enlisted recruits. So they were screening them for possible officer candidates, and they had about a thousand that they wanted an officer training group screen. So I was assigned as the senior member of a three officer group at Paris Island across the country in South Carolina. To do nothing all day long except examine these recruits. So those that passed, we sent to OCS, and the rest of them went on training. I did that and cleaned up the backlog and got the backlog cleaned up. I asked the commanding general if there was some job I could help him with, because you know we were not busy. So I was made the assistant G3. I did that for a couple of months. I was then assigned to Quantico to be an instructor in officers' training, at basic school. I was assigned as the group leader of the weapons group. This included mortars, machine guns, artillery. And we were training second lieutenants. I did that for a year. I was then assigned to an amphibious warfare school and did that for a year. Came back to basic school where I was now the CO of the headquarters company, and the base's personnel officer; I did that for a year. I was then transferred to North Carolina at Cherry Point for duty with the Second Marine Air Wing. The Marine Corps wanted to have ground officers, experienced with the air wing and aviators experienced with the division, so four marine Lt. Cols, ground Lt. Cols, went to the air wings. Four air wing guys went to the four divisions. This is while the war is still going on. I did that for a year. I'd been going to school off and on, night school. I graduated from high school, going to night school when I was in the CCC. When I was in Hawaii before the war started, I was going to the University of Hawaii night school. And I did night school till I had enough credits over time to graduate. But in order to get a diploma in the University of Maryland, you had to spend a year there on site. So I asked to do that, and the Marine Corps let me do that. So I attended the University of Maryland for a year, did that. Then I was assigned to a terrible duty station, the Marine Corps Air Station at Kaneohe. [Laughter]

E. So what did you do there?

J. The first year I was there I was the ground G3. They had two G3s; they were the operations and training officers. One for the air wing; we had an air group there. The marine brigade had a Fourth Marine Regiment infantry, an artillery battalion, a transport battalion, a marine air group and some miscellaneous squadrons. He worried about their training. I worried about the ground guys training. I did that for a year; then I was transferred to be the executive officer of the Fourth Marine Regiment. I did that for a year. During that year we had one interesting experience. This is in the 60s now. Excuse me, late '59, and the middle Asian war was warming up. And in Vietnam, the Marine Corps wanted to make a landing, loaded the brigade up, told everyone we were going to Camp Pendleton, for maneuvers. This time we went over to Lolo Lang to get ammunition before we started. And we went off well, um, west. We got off the coast of Vietnam, one day out and decided we weren't going to land. So up to this time we were telling the troops what we were going to do. Boy were they excited.

We were to do what we were supposed to do. They got the bayonets out, sharpened the bayonets, and oiled the guns and cleaning their gear. When we told the troops we were not going in, morale went like this [dipping motion with his arms], just visible. Instead we went back to Okinawa, spent five days there on rest and recreation such as it was.

E. What was the reason?

J. Ahh some political reason. And then we did send one of the battalion to the States as a maneuver enemy. The rest of it went back to Kaneohe. After a year at that job I was assigned to the commanding officer of the 1st Battalion 4th Marines. I spent almost a year there, and then in April of 1962, I retired from there, and came back to the States. When I was there, I met some civilians at a club, a civilian club, and one of them worked for IBM. He asked me what I would do after I got out of the Marine Corps. I told him I don't know, some middle management job probably. Well he said, "You should probably go talk to our guys in our office in downtown Honolulu," and I did that. When I was finally discharged, I had some interviews with some Aerospace people because I thought that would be the place. And I really didn't like their style of business. So I interviewed with the local IBM people here in San Francisco, and they accepted me in. I went to IBM training. Now the people in my IBM training class were MBAs from Stanford and Cal. That was the level of people that they were accepting. I was very fortunate to get in. It was a good company; they trained a lot. I first became a systems engineer, then I found out that the salesmen made all the money. So I shifted and became a salesman, which was true.

E. So all that training the Marines provided you in telecommunications and in radio, did that all help you?

J. Well in a way it helped me. It made me more alert to systems in general. Part of the IBMs salesmen or SE is to figure out how you can use IBM equipment in automating systems. For example, one of my jobs was to automate, at that time, credit records of the credit bureau. It was in downtown San Francisco at the time. And it was all card files, so we automated them. But that's the kind of thing you did. My Marine Corps training was good in many, many respects. It served me well.

E. Was that when IBM used punch cards?

J. Well I started off with punch cards.

E. I remember those.

J Yeah, yeah well I remember in about 1959, all the officers in the brigade were called into the movie theater for a presentation from a Lt.Col, from headquarters for our new personnel recording system. He said we are now automated, and this is the way it's going to work. Up until this time, each company filled out a morning report on who's present and what the status of every person in the unit is. And that goes to battalion and that gets transferred to a report that goes to regiment, then to division. Instead of doing that, we would fill out a record that would go to punch cards. And we will punch these cards and send them back to you and see if we did them right. I thought I don't know how we would save any work here. And that was my introduction to computing. And they did it.

E. How long were you at IBM?

J. I was there from '62 to '69, 7 years. Not very long, but it was long enough to work my way up to be a mid-level manager. At one point I was the systems engineer manager for one of the branch offices. And I was a sales manager for a large insurance group nationwide.

E. Have you always stayed in the San Francisco area since then?

A Pearl Harbor Survivors Story

J. Yes. Well IBM sent me down to LA for a year, when I was a systems engineer. I was working on the Bell Telephone account.

E. After leaving IBM, I read about your career, wow, and accomplishments.

J. I had a lot of luck and a lot of people helping. I saw an opportunity when I was at Fireman's Fund, an insurance company. And they had a large data processing section, and they had heavy turnover. They would bring in recruits, train them, and then they would go out. They were well trained, and trained programmers were in demand in other places. What I told the president, what we should do is form a company inside the Fireman's Fund Group and spin off the data processing section and have that as a company. And that way you can offer stock to key employees, you can offer benefits, which you can't offer in the insurance company. He agreed, and he provided me with a substantial amount of money to form a company. We were ready to form this company, but I had to be approved by Fireman's Fund corporate. And the president of the insurance company said no, we are not going to do that. Fortunately when I did this, I put in a provision that if the Fireman's Fund didn't do this, then I had ownership of the project, then I was able then to form the first company. And the first project we had was to install a large software system at Fireman's Fund. Our products was selling software products. Computers were just then becoming better known.

E. Could you share with us, your involvement with the Korean War memorial?

J. Back in 2006, or so the Marine [can't understand what he is saying something about a club] started having lunches for Korean War veterans. They were also having events for the Guadalcanal veterans and the Iwo Jima veterans. The wanted to recognize the Korean War veterans as well. We had a committee working with them to find the veterans and tell them about it. And I was part of that. One of the Korean War veterans, who was also a part of this group was Adm. Russell Gormin. Now deceased, but he and other Korean vets, would meet after these luncheons and talk about what we would do next year. In one of these conversations in 2009, we said we ought to have a memorial here. There is no memorial in the west. No significant one in the west coast. So we formed a committee, about that time Adm Gormin became ill and had to drop out. And a core group, several who are probably dead by now, worked to find a location. Along with this luncheon that we had, a Korean businessman and restaurant owner, Menja Kim has been having appreciation dinners for Korean Veterans. Small groups you know, 20 to 25, so we enlisted him in to this processes. So that we could get Korean input. So we had, by May 29th, we had a committee, which had Menja Kim and he offered his offices for a place to, you know, work. We hired a person to build a website. You know to do that kind of work and you know we were on the way. By 2010 Adm Gormin is gone. Myself, Menja Kim, and Don Reed had done all the dirty work to form a 401c 3. And we registered that in 2010 and now we were a formal organization. General Miat, who is on our board, said we need to get a figure head for this organization. Someone who is well known and well respected. He suggested Pete McClusty. Well, Pete McClusty is a friend of mine fortunately. So I talked to Pete and he became our first president. We were doing two things, raising money and looking for a site. Pete is a good money raiser. We were very fortunate that he knew someone at Pectal. He talked to that person and I talked to that person. And then we had some correspondence and they finally made a donation of $400,000. In the meantime, at a board meeting, Menja Kim has committed $100,000. Don Reed has committed $100,000. I committed $50,000. We had the nucleus of a building fund. We figured that we would need about three and a half million to do what we wanted to do. Lt. Lopez, I'm going to jump back a bit. He received the Medal of Honor. The basic school at Quantico, three or four years ago decided that they would create a Lopez award. And it's a statute about the size of that picture over there [points to his left] of him going over the top. I am sure you saw it in my office. It's a guy going over a ladder over the sea wall. So a sculptor in Texas, made such a sculpture and that became the Lopez award. I was invited back to Quantico about the time they awarded the first award. And I thought that would make a fine statue or memorial. As it turned out, it expanded to three panels, and we could not get the sculptor or his wife to do what we wanted to with the people in the boat. So we severed our connection and the Presidio Historical landscape architect submitted a second design and that's what you see today. He did a great job. And we spent some money with the first architect but it just didn't work out. We are friends, but this turned out to be a better depiction.

E. So through your capital campaign?

J. Our capital campaign was of course to, the first was to look for a site. One of our committee members was Dr. Suol, who was one of the North Koreans who escaped from North Korea to South Korea and into the U.S. He was a retired professor from University of Colorado, but he lived here in the area. He was very friendly with the Counsel General. He wanted it in Golden Gate Park and he thought the Korean government would fund it if we put it in Golden Gate Park. Well that wasn't a good idea for a lot of reasons. So we had a parting of the ways and he left us. Pete McClusty had a relationship with the then Superintendent Greg Middleton of the Presidio Trust. And he got us a hearing with the officers of the Presidio Trust. And we came out made a presentation, but I'm sure that Pete's relationship, this guy had worked for him when Pete was running for something. This had a lot to do with them deciding to give us space. In the meantime, Don Reed has talked to Diane Feinstein and Charlotte Shultz so they also had some influence. So you will hear two different versions of that. Mike Bounine, the planning officer was told to find four possible sites and have the committee look at them. So he gave us four sites. One was a site that we selected. One was a site at the head of the parade ground. One was a site over the tunnel and one was a site out where the monument is on the west side. You know that monument? So we examined all of these locations. By far the one across from the cemetery was hands down. So we selected that and with a lot of meetings after that, that's where we went. We were meeting with the Presidio every other week for several years and developing this jointly. Uhhmm where are we?

E. The Presidio Trust, is that a city government? Or is that the land, city land?

J. It's the National Park

E. It's the National Park.

J. However, the Presidio Trust has control of all the Presidio grounds. Congress set up a special trust and carved it out of the National Park for them. So the National Park has no say.

E. Isn't there a fort?

J Oh, there's a fort out at the end of the bridge. Fort Point but that's before the Civil War. That's part of the Presidio, however.

[LT.Col%20Stevens%20final%20(1).docx]

Biographical Notes

USMC IN THE PACIFIC: Biography Of Lieutenant Colonel John Stevens USMC retired

On 1 May 1962, LtCol John R. Stevens retired from the Marine Corps after nearly 23 years of service to the Corps.

John Stevens was born on April 22, 1921 in Butte , Montana .

LtCol Stevens enlisted in the Marine Corps in 1939. After completing boot training at the Recruit Depot in San Diego , and attending Field Telephone School at the Signal Detachment, San Diego , he was assigned to the 1st Defense Battalion. The 1st Defense Battalion was moved to Pearl Harbor in February 1941 aboard the USS Enterprise. Detachments of the 1st Defense Battalion were sent to Wake Island, Palmyra Island and Johnson Island . Late in 1941, LtCol Stevens, then Sgt, was among a group from the 1st Defense Battalion who were sent to Midway on a relief mission, returning shortly before the attack on Pearl Harbor on 7 December 1941. Shortly after Pearl Harbor, the remnants of the 1st Defense Battalion that were still at Pearl Harbor were moved to Palmyra Island . LtCol Stevens was field commissioned a second lieutenant, from Master Technical Sergeant, in August 1942, on Palmyra . Subsequent to that he was sent back to Midway to act as the communications officer for Sand Island .

Lt Stevens returned to the U.S. in July 1943. After a short tour as a coding officer for Gen HM Smith., at Camp Elliot , he was assigned to Fort Monmouth , New Jersey for further training. This was followed by a number of other training assignments before being assigned to the 1st Marine Division in 1944 as executive officer of the 1st Signal Company, on Pavuvu Island , Russell Island group. While in this capacity, LtCol Stevens, then Capt, participated in the assault and subsequent occupation of Okinawa .

From August 1945 until June 1946, Capt Stevens served as the communications officer of the 7th Marines during the occupation of North China, based at Peitaiho Beach , North China . He returned to the U.S. in June 1946. After a tour as communications officer at the Marine Corps Recruit Depot in San Diego , then Capt Stevens was sent to a one year tour at the electronic engineering school in Great Lakes, after which LtCol Stevens joined the 1st Marine Division at Camp Pendleton . He was assigned as the commanding officer, A Company, 1st Battalion, 5th Marines in 1949 and took that unit to Korea as part of the 1st Provisional Marine Brigade in July 1950. While in Korea he participated in the Pusan Perimeter campaign, the Inchon landing and capture of Seoul , the Wonsan Landing, and the Chosin Reservoir Campaign up to the end of November 1950.

Returning to the United States , LtCol Stevens was assigned to various tours, including Commanding Officer, Marine Barracks Naval Supply Center , Oakland ; as assistant G3, Parris Island Recruit Training Depot, Senior Instructor Weapons Training Group Basic School, Quantico ; Student Junior Amphibious Warfare School ; and Commanding Officer Headquarters Battalion, Basic School . This was followed by a tour at the University of Maryland , where he received a BS Degree in Military Science.

Subsequent to that, LtCol Stevens was assigned, under the cross training program, to the 2nd Marine Air-Wing at Cherry Point, N.C., where he served in various general staff assignments for two years. Following that assignment, LtCol Stevens joined the 1st Marine Brigade in Kaneohe , Hawaii in July 1959, as assistant G3.

From June 1960 to July 1961, he was Executive Officer of the 4th Marine Regiment, Reinforced and in July 1961 until retirement, was assigned as the Commanding Officer, 1st Battalion, 4th Marine Regiment, 1st Marine Brigade.

Following his retirement, he joined IBM where he served in various systems engineering and marketing management positions.

He left IBM in 1969 to found a software consulting company specializing in the insurance industry. In 1980 that company entered in to a joint venture with Quotron Systems and the Continental Corporation to market turnkey hardware/software computer systems to the independent insurance agency market place. He left the active management of that company (Insurnet) to co-found a new telecommunications management company in 1983. That company became public in 1987. Leaving the active management of that company in 1986, he founded an information management company, Stellar Net, where he served as the Chief Executive Officer until 2001.

From USS San Francisco Memorial Foundation website
[https://usssanfrancisco.org/ltcol-john-r-stevens-usmc-ret/]

Business executive positions held by John Stevens

Chairman and Founder 1969-Current American Information Development, Inc.

Chairman 1971-1974 Life Equity Information, Inc.

President 1980-1983 Insurnet, Inc.

Vice Chairman 1983-1985 Insurnet, Inc.

Co-founder,Chairman, CEO and President 1983-1985 Centex Telemanagement, Inc.

Vice Chairman 1985-1989 Centex Telemanagement, Inc.

Founder and Chairman 1987-2001 Stellarnet, Inc

Chairman 1986-2001 Industrial Work Hardening Center

Other Positions:

Past President Pacific Heights Homeowners Association.

Vice President and Founder Honolulu Civitan Club

Past President Northern California Chapter 1st Marine Division Association

Past Vice President 1st Marine Division Association (National)

Past Deputy Vice President, West, 1st Marine Division Association

Past Board Member Marine Memorial Association, San Francisco CA

Past Board Member VantageMed Corporation, Sacramento Ca

Past Board Member Collimated Holes, Los Gatos Ca

Founding President of the Chosin Few Golden Gate Chapter

Member, Marine Corps Coordinating Council of San Francisco

Board member, USS San Francisco Memorial Foundation

John Stevens's awards

2 Bronze Star Medals with Combat Vs

Combat Action Ribbon

PUC with 2 Stars

Good Conduct Medal

China Service Medal

American Defense Medal

Asiatic-Pacific Campaign Medal/ 2 stars

American Campaign Medal

Victory Medal WWII

National Defense Service Medal

Korean Service Medal with 3 Stars

UN Service Medal

Korean PUC

From USS San Francisco Memorial Foundation website
[https://usssanfrancisco.org/ltcol-john-r-stevens-usmc-ret/]

A Pearl Harbor Survivors Story

.

Photo 1: John Stevens during his boot camp graduation ceremony in 1939.
Photo credit: John Stevens

Photo 2: Captain Stevens (standing on far right) and his men at the Battle of Okinawa. Photo 3: US Marines in combat uniforms. Photo credits: John Stevens

U.S. Marine officers chart the drive to Seoul, Korea with the aid of a South Korean interpreter.

Photo 4: Captain Stevens (center pointing to map), with his platoon leaders and Korean interpreter, on the outskirts of Seoul. Photo credit: US Marine Archives

Photo 5: Lt. Col. John R. Stevens just before his retirement from the Marine Corps in 1962. Photo credit: US Marine Archives

John Stevens

Photos 6,7: Every year, Korean War and World War II veteran John Stevens marked the anniversary of the start of the Korean War by giving a speech at the Korean War Memorial in the Presidio of San Francisco. Photo (6) Michael Macor, (7) Brant Ward (SF Chronicle)

Photo 8: Lt. Col. John R. Stevens earned a Bronze Star in the Korean War.

Photo 9: Jodi and John Stevens as Chairs of Honor at the USS San Francisco Memorial.

Chairs of Honor
Jody and John Stevens

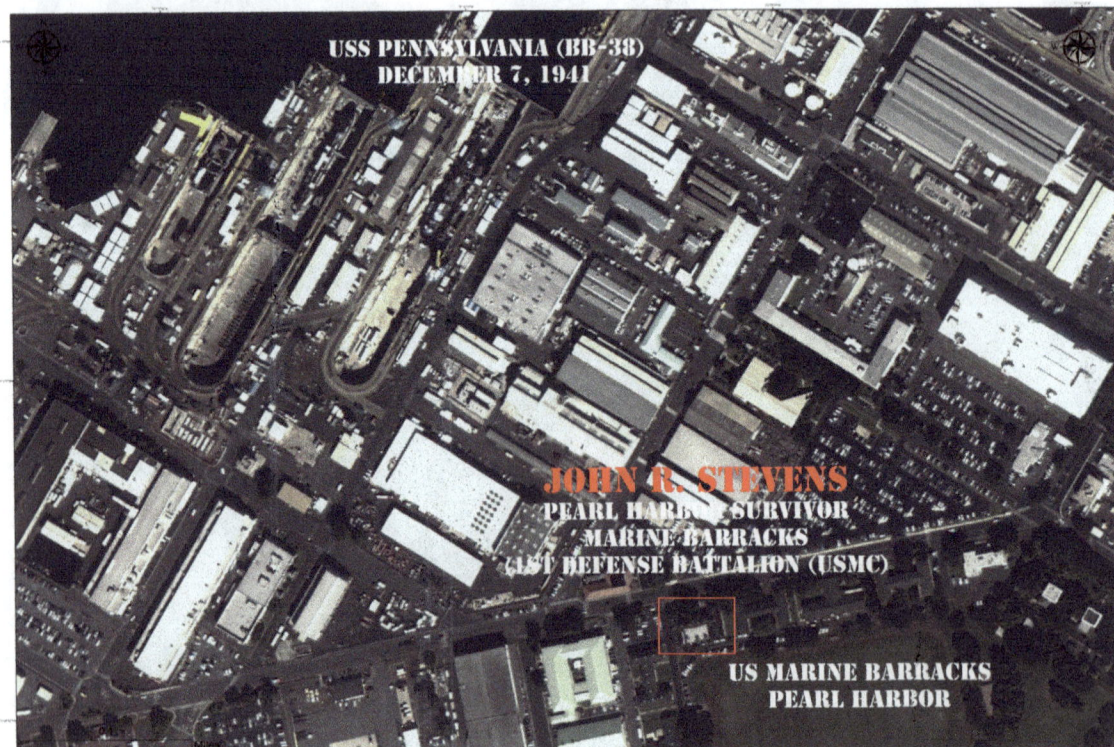

Photos 10,11: Map of John's location Marine Barracks, Pearl Harbor, December 7, 1941.

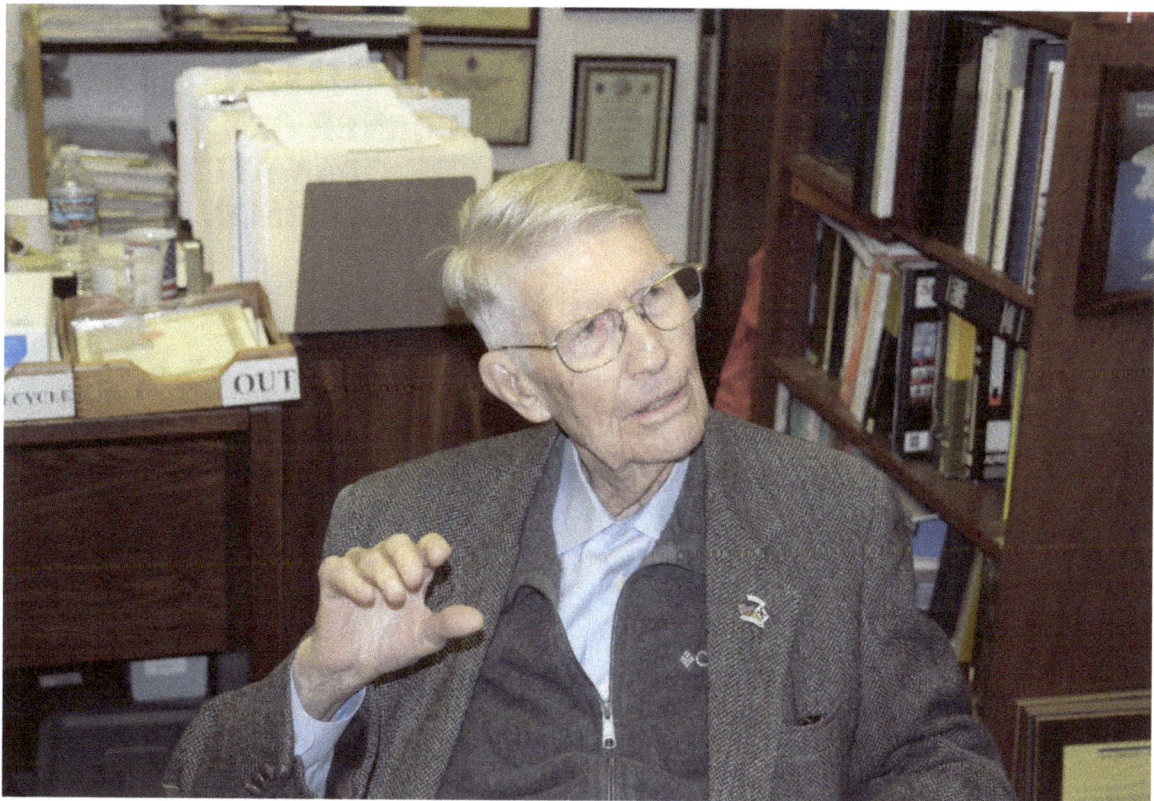

Photos 12, 13: John at his Van Ness office.

THE KOREAN WAR MEMORIAL FOUNDATION
WAS FOUNDED IN SAN FRANCISCO IN JUNE, 2010
WITH THE PURPOSE OF BUILDING A PERMANENT MEMORIAL TO COMMEMORATE
THE 60TH ANNIVERSARY OF THE KOREAN WAR.

THIS MEMORIAL WILL STAND AS A FITTING TESTAMENT TO THE MEMORY
OF THE NEARLY TWO MILLION UNITED NATIONS SERVICE MEN AND WOMEN
FROM TWENTY-ONE COUNTRIES WHO FOUGHT TO PROTECT
SOUTH KOREA'S FREEDOM DURING THE KOREAN WAR.
IT WILL SYMBOLIZE, FOR FUTURE GENERATIONS,
THE SACRED MEMORY OF THOSE WHO WENT BEFORE,
AND THE SACRIFICES THEY MADE FOR US AND FOR
FREEDOM-LOVING PEOPLE EVERYWHERE.

IN ADDITION, IT WILL SERVE TO STRENGTHEN
THE CLOSE POLITICAL AND COMMERCIAL TIES
THAT CONNECT THE UNITED STATES
AND THE REPUBLIC OF KOREA.
IT IS FITTING THAT THE MEMORIAL SITE
LOOKS WESTWARD ACROSS THE PACIFIC OCEAN,
THE WATERS THAT CONNECT OUR TWO NATIONS.

THE KOREAN WAR MEMORIAL SITE
IS SITUATED AT THE CORNER OF
LINCOLN BOULEVARD AND SHERIDAN AVENUE
IN THE PRESIDIO OF SAN FRANCISCO.

TWO STONE WALLS WILL
FLANK THE SHERIDAN AVENUE
ENTRANCE TO THE MEMORIAL PLAZA.
INSCRIBED IN THESE WALLS WILL BE
THE NAME OF THE MEMORIAL AND
THE FLAGS OF THE UNITED STATES OF AMERICA
AND THE REPUBLIC OF KOREA.

**KOREAN WAR
MEMORIAL FOUNDATION**

www.ingramcontent.com/pod-product-compliance
Lightning Source LLC
Chambersburg PA
CBHW081544040426
42448CB00015B/3221